The Nature of Space and Time

THE ISAAC NEWTON INSTITUTE
SERIES OF LECTURES

The Nature of Space and Time

Stephen Hawking and Roger Penrose

PRINCETON UNIVERSITY PRESS

PRINCETON, NEW JERSEY

Library of Congress Cataloging-in-Publication Data

Hawking, S. W. (Stephen W.)
The nature of space and time / Stephen Hawking and Roger Penrose.
p. cm. -- (Isaac Newton Institute series of lectures)
Includes bibliographic references (p.   ).
ISBN 0-691-03791-4 (cl : alk. paper)
1. Space and time. 2. Quantum theory. 3. Astrophysics.
4. Cosmology. I. Penrose, Roger. II. Title. III. Series.
QC173.59.S65HA   1995
530.1'1--dc20   95-35582

This book has been composed in Palatino

Princeton University Press books are printed on acid-free paper
and meet the guidelines for permanence and durability of the
Committee on Production Guidelines for Book Longevity of the
Council on Library Resources

http://pup.princeton.edu

Printed in the United States of America

10 9 8 7 6 5

Contents

Foreword

THE DEBATE between Roger Penrose and Stephen Hawking record-
ed in this book was the high point of a six-month program held in 1994
at the Isaac Newton Institute for Mathematical Sciences at the University of Cambridge. It represents a serious discussion of some of
the most fundamental ideas about the nature of the universe. Need-
less to say, we are not yet at the end of the road; uncertainties and
controversy still persist and there is plenty to argue about.

Some sixty years ago there was a famous and extended debate
between Niels Bohr and Albert Einstein about the foundations of
quantum mechanics. Einstein refused to accept that quantum mechanics was a final theory. He found it philosophically inadequate,
and he fought a hard battle against the orthodox interpretation of the
Copenhagen School which Bohr represented.

In a sense the debate between Penrose and Hawking is a continua-
tion of that earlier argument, with Penrose playing the role of Einstein
and Hawking that of Bohr. The issues now are more complex and
broader but as before they represent a combination of technical arguments and philosophical standpoints.

Quantum theory, or its more sophisticated version quantum field
theory, is now highly developed and technically successful, even if
there are still philosophical skeptics such as Roger Penrose. General
relativity, Einstein's theory of gravity, has equally stood the test of
time and can claim remarkable successes, although there are serious
problems concerning the role of singularities or black holes.

The real issue that dominates the Hawking-Penrose discussion is
the combination of these two successful theories, how to produce a
theory of "quantum gravity." There are deep conceptual and technical problems involved, and these provide scope for the arguments
discussed in these lectures.

Examples of the fundamental questions raised include "the arrow of time," the initial conditions at the birth of the universe, and the way that black holes swallow up information. On all these, and many others, Hawking and Penrose stake out subtly different positions. The arguments are carefully presented in both mathematical and physical terms, and the format of the debate allows for a meaningful exchange of criticism.

Although some of the presentation requires a technical understanding of the mathematics and physics, much of the argument is conducted at a higher (or deeper) level that will interest a broader audience. The reader will at least get an indication of the scope and subtlety of the ideas being discussed and of the enormous challenge of producing a coherent picture of the universe that takes full account of both gravitation and quantum theory.

Michael Atiyah

Acknowledgments

THE AUTHORS, the publisher, and the Isaac Newton Institute for Mathematical Sciences wish to extend their grateful thanks to the following individuals who assisted in the preparation of the lecture series and the book: Matthias R. Gaberdiel, Simon Gill, Jonathan B. Rogers, Daniel R. D. Scott, and Paul A. Shah.

The Nature of Space and Time

CHAPTER ONE

Classical Theory

S. W. Hawking

IN THESE LECTURES, Roger Penrose and I will put forward our related but rather different viewpoints on the nature of space and time. We shall speak alternately and shall give three lectures each, followed by a discussion on our different approaches. I should emphasize that these will be technical lectures. We shall assume a basic knowledge of general relativity and quantum theory.

There is a short article by Richard Feynman describing his experiences at a conference on general relativity. I think it was the Warsaw conference in 1962. It commented very unfavorably on the general competence of the people there and the relevance of what they were doing. That general relativity soon acquired a much better reputation, and more interest, is in considerable measure due to Roger's work. Up to then, general relativity had been formulated as a messy set of partial differential equations in a single coordinate system. People were so pleased when they found a solution that they didn't care that it probably had no physical significance. However, Roger brought in modern concepts like spinors and global methods. He was the first to show that one could discover general properties without solving the equations exactly. It was his first singularity theorem that introduced me to the study of causal structure and inspired my classical work on singularities and black holes.

I think Roger and I pretty much agree on the classical work. However, we differ in our approach to quantum gravity and indeed to quantum theory itself. Although I'm regarded as a dangerous radical by particle physicists for proposing that there may be loss of quantum coherence, I'm definitely a conservative compared to Roger. I take the positivist viewpoint that a physical theory is just a mathematical model and that it is meaningless to ask whether it corresponds to reality. All that one can ask is that its predictions should be in agreement with observation. I think Roger is a Platonist at heart but he must answer for himself.

Although there have been suggestions that spacetime may have a discrete structure, I see no reason to abandon the continuum theories that have been so successful. General relativity is a beautiful theory that agrees with every observation that has been made. It may require modifications on the Planck scale, but I don't think that will affect many of the predictions that can be obtained from it. It may be only a low energy approximation to some more fundamental theory, like string theory, but I think string theory has been oversold. First of all, it is not clear that general relativity, when combined with various other fields in a supergravity theory, cannot give a sensible quantum theory. Reports of the death of supergravity are exaggerations. One year everyone believed that supergravity was finite. The next year the fashion changed and everyone said that supergravity was bound to have divergences even though none had actually been found. My second reason for not discussing string theory is that it has not made any testable predictions. By contrast, the straightforward application of quantum theory to general relativity, which I will be talking about, has already made two testable predictions. One of these predictions, the development of small perturbations during inflation, seems to be confirmed by recent observations of fluctuations in the microwave background. The other prediction, that black holes should radiate thermally, is testable in principle. All we have to do is find a primordial black hole. Unfortunately, there don't seem to be many around in this neck of the woods. If there had been, we would know how to quantize gravity.

Neither of these predictions will be changed even if string theory is the ultimate theory of nature. But string theory, at least at its current state of development, is quite incapable of making these predictions except by appealing to general relativity as the low energy effective theory. I suspect this may always be the case and that there may not be any observable predictions of string theory that cannot also be predicted from general relativity or supergravity. If this is true, it raises the question of whether string theory is a genuine scientific theory. Is mathematical beauty and completeness enough in the absence of distinctive observationally tested predictions? Not that string theory in its present form is either beautiful or complete.

For these reasons, I shall talk about general relativity in these lectures. I shall concentrate on two areas where gravity seems to lead to features that are completely different from other field theories. The first is the idea that gravity should cause spacetime to have a beginning and maybe an end. The second is the discovery that there seems to be intrinsic gravitational entropy that is not the result of coarse graining. Some people have claimed that these predictions are only artifacts of the semiclassical approximation. They say that string theory, the true quantum theory of gravity, will smear out the singularities and will introduce correlations in the radiation from black holes so that it is only approximately thermal in the coarse-grained sense. It would be rather boring if this were the case. Gravity would be just like any other field. But I believe it is distinctively different, because it shapes the arena in which it acts, unlike other fields which act in a fixed spacetime background. It is this that leads to the possibility of time having a beginning. It also leads to regions of the universe that one can't observe, which in turn gives rise to the concept of gravitational entropy as a measure of what we can't know.

In this lecture I shall review the work in classical general relativity that leads to these ideas. In my second and third lectures (Chapters 3 and 5) I shall show how they are changed and extended when one goes to quantum theory. My second lecture will be about black holes, and the third will be on quantum cosmology.

The crucial technique for investigating singularities and black holes that was introduced by Roger, and which I helped develop, was the study of the global causal structure of spacetime. Define $I^+(p)$ to be the set of all points of the spacetime M that can be reached from p by future-directed timelike curves (see fig. 1.1). One can think of $I^+(p)$ as the set of all events that can be influenced by what happens at p. There are similar definitions in which plus is replaced by minus and future by past. I shall regard such definitions as self-evident.

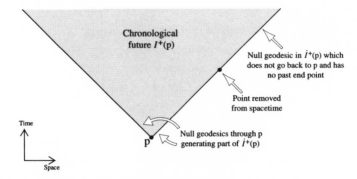

Chronological
future $I^+(p)$

Null geodesic in $\dot{I}^+(p)$ which
does not go back to p and has
no past end point

Point removed
from spacetime

Time

Null geodesics through p
generating part of $\dot{I}^+(p)$

p

Space

Figure 1.1 The chronological future of a point p.

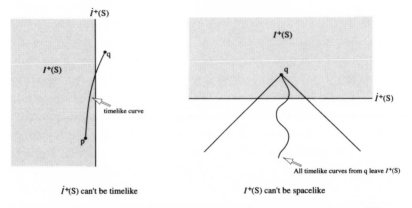

$\dot{I}^+(S)$

q

$I^+(S)$

timelike curve

p

$I^+(S)$

q

$\dot{I}^+(S)$

All timelike curves from q leave $I^+(S)$

$\dot{I}^+(S)$ can't be timelike

$I^+(S)$ can't be spacelike

Figure 1.2 The boundary of the chronological future cannot be timelike or spacelike.

One now considers the boundary $\dot{I}^+(S)$ of the future of a set S. It is fairly easy to see that this boundary cannot be timelike. For in that case, a point q just outside the boundary would be to the future of a point p just inside. Nor can the boundary of the future be spacelike, except at the set S itself. For in that case every past-directed curve from a point q, just to the future of the boundary, would cross the boundary and leave the future of S. That would be a contradiction with the fact that q is in the future of S (fig. 1.2).

One therefore concludes that the boundary of the future is null apart from the set S itself. More precisely, if q is in the boundary of the future but is not in the closure of S, there is a past-directed null

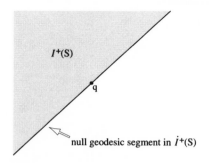

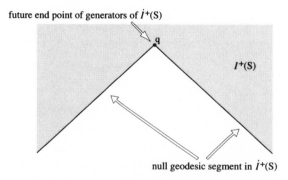

Figure 1.3 *Top*: The point q lies in the boundary of the future, so there is a null geodesic segment in the boundary which passes through q. *Bottom*: If there is more than one such segment, the point q will be their future endpoint.

geodesic segment through q lying in the boundary (see fig. 1.3). There may be more than one null geodesic segment through q lying in the boundary, but in that case q will be a future endpoint of the segments. In other words, the boundary of the future of S is generated by null geodesics that have a future endpoint in the boundary and pass into the interior of the future if they intersect another generator. On the other hand, the null geodesic generators can have past endpoints only on S. It is possible, however, to have spacetimes in which there are generators of the boundary of the future of a set S that never intersect S. Such generators can have no past endpoint.

A simple example of this is Minkowski space with a horizontal line segment removed (see fig. 1.4). If the set S lies to the past of the

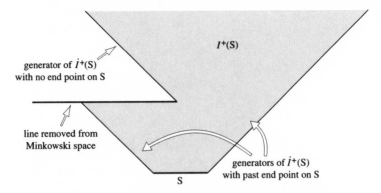

Figure 1.4 As a line has been removed from Minkowski space, the boundary of the future of the set S has a generator with no past endpoint.

horizontal line, the line will cast a shadow and there will be points just to the future of the line that are not in the future of S. There will be a generator of the boundary of the future of S that goes back to the end of the horizontal line. However, as the endpoint of the horizontal line has been removed from spacetime, this generator of the boundary will have no past endpoint. This spacetime is incomplete, but one can cure this by multiplying the metric by a suitable conformal factor near the end of the horizontal line. Although spaces like this are very artificial, they are important in showing how careful you have to be in the study of causal structure. In fact, Roger Penrose, who was one of my Ph.D. examiners, pointed out that a space like the one I just described was a counterexample to some of the claims I made in my thesis.

To show that each generator of the boundary of the future has a past endpoint on the set, one has to impose some global condition on the causal structure. The strongest and physically most important condition is that of global hyperbolicity. An open set U is said to be globally hyperbolic if

1. For every pair of points p and q in U the intersection of the future of p and the past of q has compact closure. In other words, it is a bounded diamond shaped region (fig. 1.5).
2. Strong causality holds on U. That is there are no closed or almost closed timelike curves contained in U.

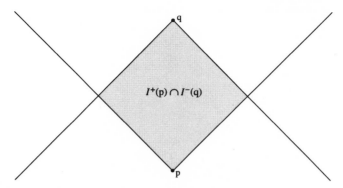

Figure 1.5 The intersection of the past of p and the future of q has compact closure.

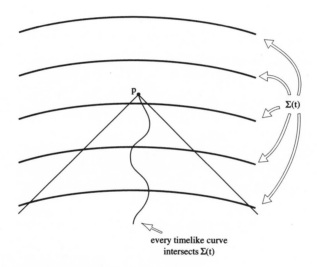

Figure 1.6 A family of Cauchy surfaces for U.

The physical significance of global hyperbolicity comes from the fact that it implies that there is a family of Cauchy surfaces $\Sigma(t)$ for U (see fig. 1.6). A Cauchy surface for U is a spacelike or null surface that intersects every timelike curve in U once and once only. One can predict what will happen in U from data on the Cauchy surface, and one can formulate a well-behaved quantum field theory on a globally hyperbolic background. Whether one can formulate a sensible

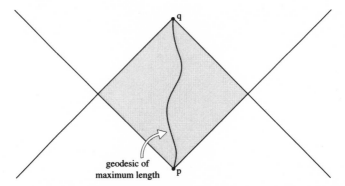

Figure 1.7 In a globally hyperbolic space, there is a geodesic of maximum length joining any pair of points that can be joined by a timelike or null curve.

quantum field theory on a nonglobally hyperbolic background is less clear. So global hyperbolicity may be a physical necessity. But my viewpoint is that one shouldn't assume it because that may be ruling out something that gravity is trying to tell us. Rather, one should deduce that certain regions of spacetime are globally hyperbolic from other physically reasonable assumptions.

The significance of global hyperbolicity for singularity theorems stems from the following. Let U be globally hyperbolic and let p and q be points of U that can be joined by a timelike or null curve. Then there is a timelike or null geodesic between p and q which maximizes the length of timelike or null curves from p to q (fig. 1.7). The method of proof is to show that the space of all timelike or null curves from p to q is compact in a certain topology. One then shows that the length of the curve is an upper semicontinuous function on this space. It must therefore attain its maximum, and the curve of maximum length will be a geodesic because otherwise a small variation will give a longer curve.

One can now consider the second variation of the length of a geodesic γ. One can show that γ can be varied to a longer curve if there is an infinitesimally neighboring geodesic from p which intersects γ again at a point r between p and q. The point r is said to be conjugate to p (fig. 1.8). One can illustrate this by considering two points p and q

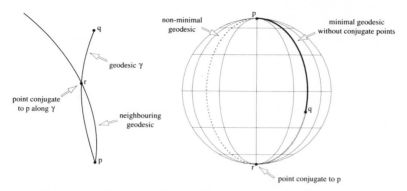

Figure 1.8 *Left*: if there is a conjugate point r between p and q on a geodesic, it is not the geodesic of minimum length. *Right*: The nonminimal geodesic from p to q has a conjugate point at the south pole.

on the surface of the Earth. Without loss of generality, one can take p to be at the north pole. Because the Earth has a positive definite metric rather than a Lorentzian one, there is a geodesic of minimal length, rather than a geodesic of maximum length. This minimal geodesic will be a line of longitude running from the north pole to the point q. But there will be another geodesic from p to q which runs down the back from the north pole to the south pole and then up to q. This geodesic contains a point conjugate to p at the south pole where all the geodesics from p intersect. Both geodesics from p to q are stationary points of the length under a small variation. But now in a positive definite metric the second variation of a geodesic containing a conjugate point can give a shorter curve from p to q. Thus, in the example of the Earth, we can deduce that the geodesic that goes down to the south pole and then comes up is not the shortest curve from p to q. This example is very obvious. However, in the case of spacetime one can show that under certain assumptions there ought to be a globally hyperbolic region in which there should be conjugate points on every geodesic between two points. This establishes a contradiction which shows that the assumption of geodesic completeness, which can be taken as a definition of a nonsingular spacetime, is false.

The reason one gets conjugate points in spacetime is that gravity is an attractive force. It therefore curves spacetime in such a way that neighboring geodesics are bent toward each other rather than away. One can see this from the Raychaudhuri or Newman-Penrose equation, which I will write in a unified form.

Raychaudhuri-Newman-Penrose Equation

$$\frac{d\rho}{dv} = \rho^2 + \sigma^{ij}\sigma_{ij} + \frac{1}{n}R_{ab}l^a l^b,$$

where $n = 2$ for null geodesics,
$n = 3$ for timelike geodesics.

Here v is an affine parameter along a congruence of geodesics with tangent vector l^a which is hypersurface orthogonal. The quantity ρ is the average rate of convergence of the geodesics, while σ measures the shear. The term $R_{ab}l^a l^b$ gives the direct gravitational effect of the matter on the convergence of the geodesics.

Einstein Equation

$$R_{ab} - \frac{1}{2}g_{ab}R = 8\pi T_{ab}.$$

Weak Energy Condition

$$T_{ab}v^a v^b \geq 0$$

for any timelike vector v^a.

By the Einstein equations, it will be nonnegative for any null vector l^a if the matter obeys the so-called weak energy condition. This says that the energy density T_{00} is nonnegative in any frame. The weak energy condition is obeyed by the classical energy momentum tensor of any reasonable matter, such as a scalar or electromagnetic field or a fluid with a reasonable equation of state. It may not, however, be satisfied locally by the quantum mechanical expectation value of the energy momentum tensor. This will be relevant in my second and third lectures (chapters 3 and 5).

Suppose the weak energy condition holds, and that the null geodesics from a point p begin to converge again and that ρ has the positive value ρ_0. Then the Newman-Penrose equation would imply that the convergence ρ would become infinite at a point q within an affine parameter distance $\frac{1}{\rho_0}$ if the null geodesic can be extended that far.

If $\rho = \rho_0$ at $v = v_0$ then $\rho \geq \frac{1}{\rho^{-1}+v_0-v}$. Thus there is a conjugate point before $v = v_0 + \rho^{-1}$.

Infinitesimally neighboring null geodesics from p will intersect at q. This means the point q will be conjugate to p along the null geodesic γ joining them. For points on γ beyond the conjugate point q there will be a variation of γ that gives a timelike curve from p. Thus γ cannot lie in the boundary of the future of p beyond the conjugate point q. So γ will have a future endpoint as a generator of the boundary of the future of p (fig. 1.9).

The situation with timelike geodesics is similar, except that the strong energy condition that is required to make $R_{ab}l^a l^b$ nonnegative for every timelike vector l^a is, as its name suggests, rather stronger. It is still, however, physically reasonable, at least in an averaged sense, in classical theory. If the strong energy condition holds, and the timelike geodesics from p begin converging again, then there will be a point q conjugate to p.

Strong Energy Condition

$$T_{ab}v^a v^b \geq \frac{1}{2}v^a v_a T$$

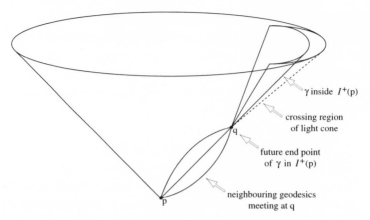

γ inside $I^+(p)$

crossing region
of light cone

future end point
of γ in $I^+(p)$

neighbouring geodesics
meeting at q

q

p

Figure 1.9 The point q is conjugate to p along null geodesics, so a null geodesic γ that joins p to q will leave the boundary of the future of p at q.

Finally, there is the generic energy condition. This says that first the strong energy condition holds. Second, every timelike or null geodesic encounters some point where there is some curvature that is not specially aligned with the geodesic. The generic energy condition is not satisfied by a number of known exact solutions. But these are rather special. One would expect it to be satisfied by a solution that was "generic" in an appropriate sense. If the generic energy condition holds, each geodesic will encounter a region of gravitational focussing. This will imply that there are pairs of conjugate points if one can extend the geodesic far enough in each direction.

The Generic Energy Condition

1. The strong energy condition holds.
2. Every timelike or null geodesic contains a point where $l_{[a}R_{b]cd[e}l_{f]}l^c l^d \neq 0$.

One normally thinks of a spacetime singularity as a region in which the curvature becomes unboundedly large. However, the trouble with that as a definition is that one could simply leave out the singular points and say that the remaining manifold was the whole of spacetime. It is therefore better to define spacetime as the maximal manifold on which the metric is suitably smooth. One can then recognize the occurrence of singularities by the existence of incomplete geodesics that cannot be extended to infinite values of the affine parameter.

Definition of Singularity

A spacetime is singular if it is timelike or null geodesically incomplete but cannot be embedded in a larger spacetime.

This definition reflects the most objectionable feature of singularities, that there can be particles whose history has a beginning or end at a finite time. There are examples in which geodesic incompleteness can occur with the curvature remaining bounded, but it is thought that generically the curvature will diverge along incomplete geodesics. This is important if one is to appeal to quantum effects to solve the problems raised by singularities in classical general relativity.

Between 1965 and 1970 Penrose and I used the techniques I have described to prove a number of singularity theorems. These theorems had three kinds of conditions. First there was an energy condition such as the weak, strong, or generic energy conditions. Then there was some global condition on the causal structure such as that there shouldn't be any closed timelike curves. And finally, there was some condition that gravity was so strong in some region that nothing could escape.

Singularity Theorems

     1. Energy condition.
     2. Condition on global structure.
     3. Gravity strong enough to trap a region.

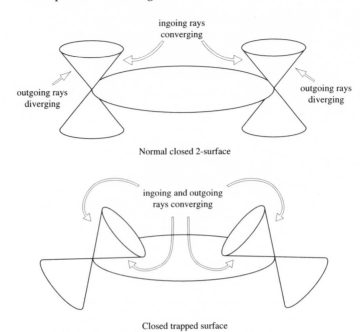

ingoing rays
converging

outgoing rays
diverging

outgoing rays
diverging

Normal closed 2-surface

ingoing and outgoing
rays converging

Closed trapped surface

Figure 1.10 At a normal closed surface, the outgoing null rays from the surface diverge, while the ingoing rays converge. On a closed trapped surface, both the ingoing and outgoing null rays converge.

This third condition could be expressed in various ways. One way would be that the spatial cross section of the universe was closed, for then there would be no outside region to escape to. Another would be that there was what was called a closed trapped surface. This is a closed two-surface such that both the ingoing and outgoing null geodesics orthogonal to it were converging (fig. 1.10). Normally if you have a spherical two-surface in Minkowski space, the ingoing null geodesics are converging but the outgoing ones are diverging. But in the collapse of a star the gravitational field can be so strong that the light cones are tipped inward. This means that even the outgoing null geodesics are converging.

The various singularity theorems show that spacetime must be timelike or null geodesically incomplete if different combinations of the three kinds of conditions hold. One can weaken one condition if

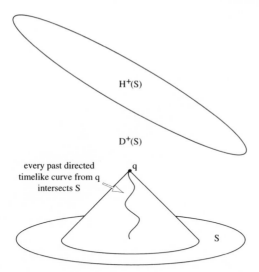

Figure 1.11 The future Cauchy development $D^+(S)$ of a set S and its future boundary, the Cauchy horizon $H^+(S)$.

one assumes stronger versions of the other two. I shall illustrate this by describing the Hawking-Penrose theorem. This has the generic energy condition, the strongest of the three energy conditions. The global condition is fairly weak, that there should be no closed timelike curves. And the no-escape condition is the most general, that there should be either a trapped surface or a closed spacelike three-surface.

For simplicity, I shall just sketch the proof for the case of a closed spacelike three-surface S. One can define the future Cauchy development $D^+(S)$ to be the region of points q from which every past-directed timelike curve intersects S (fig. 1.11). The Cauchy development is the region of spacetime that can be predicted from data on S. Now suppose that the future Cauchy development was compact. This would imply that the Cauchy development would have a future boundary called the *Cauchy horizon*, $H^+(S)$. By an argument similar to that for the boundary of the future of a point, the Cauchy horizon would be generated by null geodesic segments without past endpoints. However, since the Cauchy development is assumed to be compact, the Cauchy horizon will also be compact. This means that

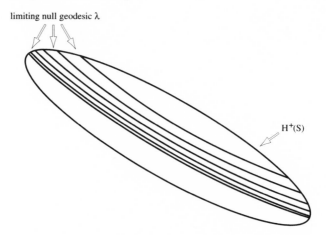

limiting null geodesic λ

H$^+$(S)

Figure 1.12 There is a limiting null geodesic λ in the Cauchy horizon which has no past or future endpoints in the Cauchy horizon.

the null geodesic generators will wind around and around inside a compact set. They will approach a limiting null geodesic λ that will have no past or future endpoints in the Cauchy horizon (fig. 1.12). But if λ were geodesically complete, the generic energy condition would imply that it would contain conjugate points p and q. Points on λ beyond p and q could be joined by a timelike curve. But this would be a contradiction because no two points of the Cauchy horizon can be timelike separated. Therefore either λ is not geodesically complete and the theorem is proved, or the future Cauchy development of S is not compact.

In the latter case one can show there is a future-directed timelike curve, γ from S, that never leaves the future Cauchy development of S. A rather similar argument shows that γ can be extended to the past to a curve that never leaves the past Cauchy development $D^-(S)$ (fig. 1.13). Now consider a sequence of points x_n on γ tending to the past and a similar sequence y_n tending to the future. For each value of n the points x_n and y_n are timelike separated and are in the globally hyperbolic Cauchy development of S. Thus, there is a timelike geodesic of maximum length λ_n from x_n to y_n. All the λ_n will cross the compact spacelike surface S. This means that there will be

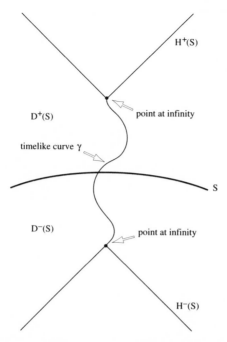

H⁺(S)

D⁺(S)   point at infinity

timelike curve γ

S

D⁻(S)   point at infinity

H⁻(S)

Figure 1.13 If the future (past) Cauchy development is not compact, there is a future (past) directed timelike curve from S that never leaves the future (past) Cauchy development.

a timelike geodesic λ in the Cauchy development which is a limit of the timelike geodesics λ_n (fig. 1.14). Either λ will be incomplete, in which case the theorem is proved, or it will contain conjugate points because of the generic energy condition. But in that case λ_n would contain conjugate points for n sufficiently large. This would be a contradiction because the λ_n are supposed to be curves of maximum length. One can therefore conclude that the spacetime is timelike or null geodesically incomplete. In other words, there is a singularity.

The theorems predict singularities in two situations. One is in the future in the gravitational collapse of stars and other massive bodies. Such singularities would be an end of time, at least for particles moving on the incomplete geodesics. The other situation in which singularities are predicted is in the past, at the beginning of the present expansion of the universe. This led to the abandonment

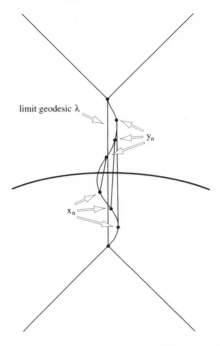

limit geodesic λ

y_n

x_n

Figure 1.14 The geodesic λ, which is the limit of the γ_n, will have to be incomplete, because otherwise it would contain conjugate points.

of attempts (mainly by the Russians) to argue that there was a previous contracting phase and a nonsingular bounce into expansion. Instead, almost everyone now believes that the universe, and time itself, had a beginning at the big bang. This is a discovery far more important than a few miscellaneous unstable particles, but not one that has been so well recognized by Nobel prizes.

The prediction of singularities means that classical general relativity is not a complete theory. Because the singular points have to be cut out of the spacetime manifold, one cannot define the field equations there and cannot predict what will come out of a singularity. With the singularity in the past the only way to deal with this problem seems to be to appeal to quantum gravity. I shall return to this in my third lecture (Chapter 5). But the singularities that are predicted in the future seem to have a property that Penrose has called *cosmic censorship*. That is, they conveniently occur in places like black holes that are hidden from external observers. So any breakdown of predictability that may occur at these singularities won't affect what happens in the outside world, at least not according to classical theory.

Cosmic Censorship

Nature abhors a naked singularity.

However, as I shall show in my next lecture, there is unpredictability in the quantum theory. This is related to the fact that gravitational fields can have intrinsic entropy that is not just the result of coarse graining. Gravitational entropy, and the fact that time has a beginning and may have an end, are the two themes of my lectures because they are the ways in which gravity is distinctly different from other physical fields.

The fact that gravity has a quantity that behaves like entropy was first noticed in the purely classical theory. It depends on Penrose's *cosmic censorship conjecture*. This is unproved, but it is believed to be true for suitably general initial data and equations of state. I shall use a weak form of cosmic censorship. One makes the approximation of treating the region around a collapsing star as asymptotically flat. Then, as Penrose showed, one can conformally embed the spacetime manifold M in a manifold with boundary $\bar{M}$ (fig 1.15). The boundary ∂M will be a null surface and will consist of two components, future and past null infinity, called $\mathcal{I}^+$ and $\mathcal{I}^-$. I shall say that weak cosmic censorship holds if two conditions are satisfied. First, it is assumed that the null geodesic generators of $\mathcal{I}^+$ are complete in a certain conformal metric. This implies that observers far from the collapse live to an old age and are not wiped out by a thunderbolt singularity sent out from the collapsing star. Second, it is assumed that the past of $\mathcal{I}^+$ is globally hyperbolic. This means there are no naked singularities that can be seen from large distances. Penrose has a stronger form of cosmic censorship, which assumes that the whole spacetime is globally hyperbolic. But the weak form will suffice for my purposes.

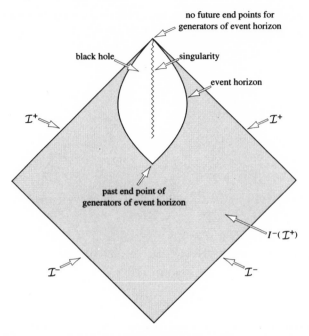

Figure 1.15 A collapsing star conformally embedded in a manifold with boundary.

Weak Cosmic Censorship

  1. $\mathcal{I}^+$ and $\mathcal{I}^-$ are complete.
  2. $I^-(\mathcal{I}^+)$ is globally hyperbolic.

If weak cosmic censorship holds, the singularities that are predicted to occur in gravitational collapse can't be visible from $\mathcal{I}^+$. This means that there must be a region of spacetime that is not in the past of $\mathcal{I}^+$. This region is said to be a black hole, because no light or anything else can escape from it to infinity. The boundary of the black hole region is called the *event horizon*. Because it is also the boundary of the past of $\mathcal{I}^+$, the event horizon will be generated by null geodesic segments that may have past endpoints but don't have any future endpoints. It then follows that if the weak energy condition holds,

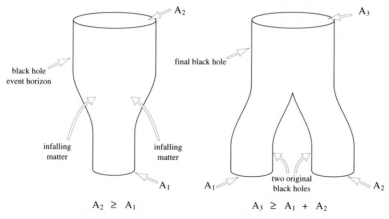

$A_2 \geq A_1$ $\qquad\qquad\qquad$ $A_3 \geq A_1 + A_2$

Figure 1.16 When we throw matter into a black hole, or allow two black holes to merge, the total area of the event horizons will never decrease.

the generators of the horizon can't be converging. For if they were, they would intersect each other within a finite distance.

This implies that the area of a cross section of the event horizon can never decrease with time, and in general will increase. Moreover, if two black holes collide and merge together, the area of the final black hole will be greater than the sum of the areas of the original black holes (fig. 1.16). This is very similar to the behavior of entropy according to the second law of thermodynamics. Entropy can never decrease and the entropy of a total system is greater than the sum of its constituent parts.

Second Law of Black Hole Mechanics

$$\delta A \geq 0$$

Second Law of Thermodynamics

$$\delta S \geq 0$$

First Law of Black Hole Mechanics

$$\delta E = \frac{\kappa}{8\pi}\delta A + \Omega\delta J + \Phi\delta Q$$

First Law of Thermodynamics

$$\delta E = T\delta S + P\delta V$$

The similarity with thermodynamics is increased by what is called the *first law of black hole mechanics*. This relates the change in mass of a black hole to the change in the area of the event horizon and the change in its angular momentum and electric charge. One can compare this to the first law of thermodynamics, which gives the change in internal energy in terms of the change in entropy and the external work done on the system. One sees that if the area of the event horizon is analogous to entropy, then the quantity analogous to temperature is what is called the surface gravity of the black hole κ. This is a measure of the strength of the gravitational field on the event horizon. The similarity with thermodynamics is further increased by the so-called *zeroth law of black hole mechanics*: the surface gravity is the same everywhere on the event horizon of a time-independent black hole.

Zeroth Law of Black Hole Mechanics

κ is the same everywhere on the horizon of a time-independent black hole.

Zeroth Law of Thermodynamics

T is the same everywhere for a system in thermal equilibrium.

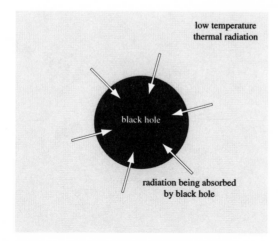

low temperature
thermal radiation

black hole

radiation being absorbed
by black hole

Figure 1.17 A black hole in contact with thermal radiation will absorb some of the radiation, but classically cannot send anything out.

Encouraged by these similarities, Bekenstein (1972) proposed that some multiple of the area of the event horizon actually was the entropy of a black hole. He suggested a generalized second law: the sum of this black hole entropy and the entropy of matter outside black holes would never decrease.

Generalized Second Law

$$\delta(S + cA) \geq 0$$

However, this proposal was not consistent. If black holes have an entropy proportional to horizon area, they should also have a nonzero temperature proportional to surface gravity. Consider a black hole that is in contact with thermal radiation at a temperature lower than the black hole temperature (fig. 1.17). The black hole will absorb some of the radiation but won't be able to send anything out, because

Figure 1.18

according to classical theory nothing can get out of a black hole. One thus has heat flow from the low-temperature thermal radiation to the higher-temperature black hole. This would violate the generalized second law because the loss of entropy from the thermal radiation would be greater than the increase in black hole entropy. However, as we shall see in my next lecture, consistency was restored when it was discovered that black holes are sending out radiation that was exactly thermal. This is too beautiful a result to be a coincidence or just an approximation. So it seems that black holes really do have intrinsic gravitational entropy. As I shall show, this is related to the nontrivial topology of a black hole. The intrinsic entropy means that gravity introduces an extra level of unpredictability over and above the uncertainty usually associated with quantum theory. So Einstein was wrong when he said, "God does not play dice." Consideration of black holes suggests, not only that God does play dice, but that he sometimes confuses us by throwing them where they can't be seen (fig. 1.18).

CHAPTER TWO

Structure of Spacetime Singularities

R. Penrose

IN THE FIRST LECTURE by Stephen Hawking, singularity theorems were discussed. The essential content of these theorems is that under reasonable (global) physical conditions, singularities must be expected. They do not say anything about the nature of the singularities, or where the singularities are to be found. On the other hand, the theorems are very general. A natural question to ask is, therefore, what the geometric nature of a spacetime singularity is. It is usually assumed that the characteristic of a singularity is that the curvature diverges. However, this is not exactly what the singularity theorems by themselves imply.

Singularities occur in the big bang, in black holes, and in the big crunch (which might be regarded as a union of black holes). They also might appear as naked singularities. Related to this question is what is called *cosmic censorship*, namely the hypothesis that these naked singularities do not occur.

To explain the idea of cosmic censorship, let me recall a bit the history of the subject. The first explicit example of a solution of Einstein's equations describing a black hole was the collapsing dust cloud of Oppenheimer and Snyder (1939). There is a singularity inside, but it is not visible from outside, as it is surrounded by the event horizon. This horizon is the surface inside of which events cannot send signals out to infinity. It was tempting to believe that this picture is generic, i.e., that it represents the general gravitational collapse. However, the OS model has a special symmetry (namely, spherical symmetry), and it is not obvious that it is really representative.

As the Einstein equations are generally hard to solve, one looks instead for global properties that imply the existence of singularities.

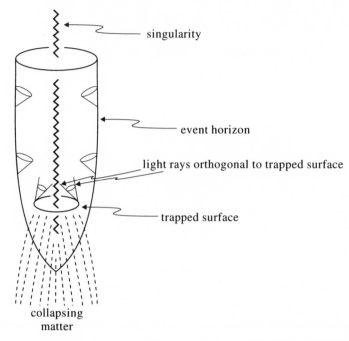

Figure 2.1 The Oppenheimer-Snyder collapsing dust cloud, illustrating a trapped surface.

For example, the OS model has a trapped surface, which is a surface whose area will decrease along light rays that are initially orthogonal to it (fig. 2.1).

One might try to show that the existence of a trapped surface implies that there is a singularity. (This was the first singularity theorem I was able to establish, on the basis of reasonable causality assumptions but no spherical symmetry being assumed; see Penrose 1965.) One can also derive similar results by assuming the existence of a converging light cone (Hawking and Penrose 1970; this occurs when all the light rays emitted in different directions from a point start to converge toward each other at a later time).

Stephen Hawking (1965) observed, very early on, that one can also turn my original argument upside down on a cosmological scale, i.e., apply it to the time-reversed situation. A reversed trapped surface then implies that there had been a singularity in the past (making appropriate causality assumptions). Now, the (time-reversed) trapped surface is very large, being on a cosmological scale.

We are here mainly concerned with analyzing the situation of a black hole. We know that there has to be a singularity somewhere, but in order to get a black hole we have to show that it is surrounded by an event horizon. The cosmic censorship hypothesis asserts just this, essentially that one cannot see the singularity itself from outside. In particular it implies that there is some region that cannot send signals to external infinity. The boundary of this region is the event horizon. We can also apply a theorem given in Stephen's last lecture to this boundary, as the event horizon is the boundary of the past of future null infinity. Thus we know that this boundary

- must be a null surface where it is smooth, generated by null geodesics,
- contains a future-endless null geodesic originating from each point at which it is not smooth,

and that

- the area of spatial cross sections cannot ever decrease with time.

It has also, in effect, been shown (Israel 1967, Carter 1971, Robinson 1975, Hawking 1972) that the asymptotic future limit of such a spacetime is the Kerr spacetime. This is a very remarkable result, as the Kerr metric is a very nice exact solution of the Einstein vacuum equations. This argument also relates to the issue of black hole entropy and I shall, in effect, come back to it in the next lecture (Chapter 4).

Accordingly, we indeed have something with a qualitative similarity to the OS solution. There are some modifications—namely, that we end up with the Kerr solution rather than the Schwarzschild solution—but these are relatively minor. The essential picture is rather similar.

However, the precise arguments are based on the cosmic censorship hypothesis. In fact, cosmic censorship is very important, as the whole theory depends upon it, and without it we might see dreadful

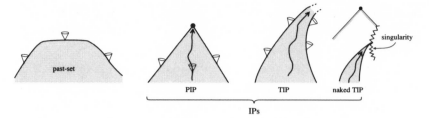

Figure 2.2 Past-sets, PIPs, and TIPs.

things instead of a black hole. So we do really have to ask ourselves whether it is true. A long time ago I thought that this hypothesis might be false and I made various attempts to find counterexamples. (Stephen Hawking once claimed that one of the strongest justifications for the cosmic censorship hypothesis was the fact that I had tried and failed to prove that it is wrong—but I think this is a very feeble argument!)

I want to discuss cosmic censorship in the context of certain ideas concerning *ideal* points for spacetimes. (These concepts are due to Seifert 1971, and Geroch, Kronheimer, and Penrose 1972). The basic idea is that one should incorporate into the spacetime actual "singular points" and "points at infinity," namely the *ideal points*. Let me first introduce the concept of an IP, i.e., an *indecomposable past-set*. Here a "past-set" is a set which contains its own past, and "indecomposable" means that it cannot be split into two past-sets neither of which contains the other. There is a theorem which tells us that one can also describe any IP as the past of some timelike curve (fig. 2.2).

There are two categories of IP, namely PIPs and TIPs. A PIP is a *proper* IP, i.e., the past of a spacetime point. A TIP is a *terminal* IP, not the past of an actual point in spacetime. TIPs define the future ideal points. Furthermore, one can distinguish TIPs according to whether this ideal point is "at infinity" (in which case there is a timelike curve generating the IP of infinite proper length)—an ∞-TIP—or a *singularity* (in which case every timelike curve generating it has finite proper length)—a singular TIP. Obviously all these concepts can be similarly applied to future-sets rather than to past-sets. In this case we have

IFs (indecomposable futures), divided into PIFs and TIFs, the TIFs being subdivided into ∞-TIFs and singular TIFs. Let me also remark that for all this to work we have to assume, in effect, that there are no closed timelike curves—actually a marginally weaker condition: no two points have the same future or the same past.

How can we describe naked singularities and the cosmic censorship hypothesis in this framework? First of all, the cosmic censorship hypothesis should not exclude the big bang (since otherwise cosmologists would be in big trouble). Now, things always come out of the big bang and never fall into it. Thus, we might try to define a naked singularity as something that a timelike curve can both enter and exit from. Then the big bang problem is automatically taken care of. It does not count as naked. In this framework we can define a *naked* TIP as a TIP that is contained in a PIP. This is an essentially local definition, i.e., we do not require the observer to be at infinity. It turns out (Penrose 1979) that the exclusion of naked TIPs is the same condition in a spacetime if we replace "past" by "future" in this definition (exclusion of naked TIFs). The hypothesis that such naked TIPs (or, equivalently, TIFs) do not occur in generic spacetimes is called the *strong cosmic censorship* hypothesis. Its intuitive meaning is that a singular point (or infinite point)—the TIP in question—cannot simply "appear" in the middle of a spacetime in such a way that it is "visible" at some finite point—the vertex of the PIP in question. It is sensible that the observer needn't be at infinity since in a given spacetime we might not know whether there actually is an infinity. Furthermore, if the strong cosmic censorship hypothesis were violated we could, at a finite time, observe a particle actually falling into a singularity, where the rules of physics would cease to hold (or else reaching infinity, which is about as bad). We can also express the *weak cosmic censorship* hypothesis in this language: we just have to replace PIP by ∞-TIP.

The strong cosmic censorship hypothesis implies that a generic spacetime with matter, subject to reasonable equations of state (for example, vacuum), can be extended to one that is free of naked singularities (naked singular TIPs). It turns out (Penrose 1979) that the exclusion of naked TIPs is equivalent to global hyperbolicity, or that the spacetime is the whole domain of dependence of some Cauchy surface (Geroch 1970). We note that this formulation of the strong cosmic censorship is manifestly symmetric in time: we can interchange future and past if we interchange IPs and IFs.

In general, we need additional conditions to rule out *thunderbolts*. By a thunderbolt we mean a singularity which reaches null infinity, destroying the spacetime as it goes (cf. Penrose 1978, fig. 7). This need not violate cosmic censorship as stated. There exist stronger versions of cosmic censorship which take care of this (Penrose 1978, condition CC4).

So let us come back to the question whether cosmic censorship is true. First of all, let us note that it is probably not true in quantum gravity. In particular, exploding black holes (about which Stephen Hawking will explain more later) result in situations where cosmic censorship seems to be violated.

In classical general relativity there are various results in both directions. In one attempt to disprove cosmic censorship I derived certain inequalities which would hold if cosmic censorship were true (Penrose 1973). In fact, they did turn out to be true (Gibbons 1972)—and this seems to give support to the idea that something like cosmic censorship should hold. On the negative side there are some special examples (which, however, violate the genericity condition) and some sketchy numerical evidence that is subject to various objections. There are, furthermore, some indications about which I have only learned very recently—in fact, Gary Horowitz only mentioned them to me yesterday—that some of the aforementioned inequalities do not hold if the cosmological constant is positive. Personally I have always believed that the cosmological constant should be zero, but it would be very interesting if cosmic censorship depended upon it being, say, nonpositive. In particular, there might be an intriguing relationship between the nature of the singularities and the nature of infinity. Infinity is spacelike if the cosmological constant is positive, but null if it is zero. Correspondingly, singularities might sometimes turn out to be timelike (which means naked, i.e., violating cosmic censorship) if the cosmological constant is positive, but perhaps singularities cannot be timelike (i.e., satisfying cosmic censorship) if it is zero.

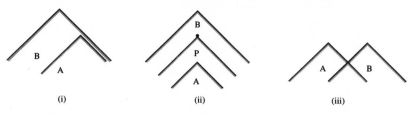

Figure 2.3 Causal relations between IPs: (i) *A* causally precedes *B*; (ii) *A* chronologically precedes *B*; (iii) *A* and *B* are spacelike separated.

To discuss the timelike or spacelike nature of singularities, let me explain the causal relations between IPs. Generalizing causality between points, we can say that an IP *A* causally precedes an IP *B*, if $A \subset B$; and *A* chronologically proceeds *B*, if there is a PIP *P* such that $A \subset P \subset B$. We call *A* and *B* spacelike separated if neither causally precedes the other (fig. 2.3).

Strong cosmic censorship can then be expressed as saying that generic singularities are never timelike. Spacelike (or null) singularities can be either of past or future type. Hence, if strong cosmic censorship holds, singularities fall into two classes:

(P) Past types, defined by TIFs.
(F) Future type, defined by TIPs.

Naked singularities could unite the two possibilities into one, as a naked singularity would be a TIP and a TIF at the same time. Therefore it is really a consequence of cosmic censorship that these classes are separate. Typical examples of class (F) are singularities in black holes and the big crunch (if it exists), and of class (P) the big bang and possibly white holes (if they exist). I do not actually believe that the big crunch is likely to happen (for ideological reasons that I shall come to in the final lecture), and white holes are very much more unlikely because they disobey the second law of thermodynamics.

Perhaps the two types of singularity satisfy completely different laws. Maybe the quantum gravity laws for them should indeed be quite different. I think that Stephen Hawking disagrees here with

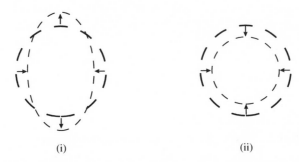

(i)   (ii)

Figure 2.4 The acceleration effects of spacetime curvature: (i) the tidal distortion due to Weyl curvature; (ii) the volume-decreasing effect of Ricci curvature.

me [SWH: "Yes!"], but I regard the following as evidence for this proposal:

(1) The second law of thermodynamics.
(2) The observations of the early universe (e.g., COBE), which indicate that it was very uniform.
(3) The existence of black holes (virtually observed).

From (1) and (2) it can be argued that the big bang singularity was extremely uniform, and from (1) that it is free of white holes (for white holes violently disobey the second law of thermodynamics). Thus, very different laws must hold for the singularities of black holes (3). To describe this difference more precisely, recall that the spacetime curvature is described by the Riemann tensor R_{abcd}, which is the sum of the Weyl tensor C_{abcd} (describing the tidal distortions, which are volume preserving to first order) and a part equivalent to the Ricci tensor R_{ab} (times the metric g_{cd}, with indices appropriately scrambled), which describes volume-decreasing distortions (fig. 2.4).

In the standard cosmological models (due to Friedmann, Lemaître, Robertson, and Walker; see, for example, Rindler 1977) the big bang has vanishing Weyl tensor. (There is also a converse to this, proved by R.P.A.C. Newman, in which a universe with an initial singularity of a conformally regular type with vanishing Weyl tensor must, if suitable equations of state hold, be an FLRW universe; see Newman

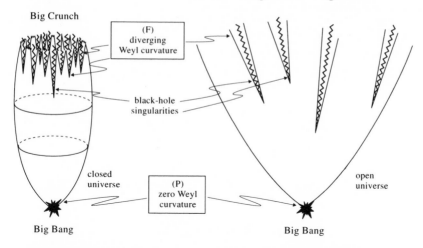

Figure 2.5 The Weyl curvature hypothesis: initial singularities (big bang) are constrained to have vanishing Weyl curvature whereas at final singularities, the Weyl curvature is expected to diverge.

1993.) On the other hand, black/white hole singularities have (in the generic case) diverging Weyl tensor. This suggests the following:

Weyl Curvature Hypothesis

- Initial-type (P) singularities are constrained to have vanishing Weyl tensor.
- Final-type (F) singularities are not constrained.

This is closely in agreement with what one sees. If the universe is closed, the final singularity (the big crunch) will have diverging Weyl tensor, in an open universe the created black holes also have diverging Weyl tensor (see fig. 2.5).

Further support for this hypothesis comes from the fact that the constraint that the early universe was fairly smooth and free of white holes reduces the phase space in the early universe by a factor of at least

$$10^{10^{123}}.$$

(This figure is the allowable phase space volume for a black hole of 10^{80} baryons, as follows from the Bekenstein-Hawking black hole entropy formula—Bekenstein 1972, Hawking 1975—and the universe has at least this much matter.)

Thus there should be a law which forces this rather unlikely result to happen! The Weyl curvature hypothesis would provide a law of this kind.

QUESTIONS AND ANSWERS

Question: Do you think that quantum gravity removes singularities?

Answer: I don't think it can be quite like that. If it were like that, the big bang would have resulted from a previously collapsing phase. We must ask how that previous phase could have had such a low entropy. This picture would sacrifice the best chance we have of explaining the second law. Moreover, the singularities of collapsing and expanding universes would have to be somehow joined together, but they seem to have very different geometries. A true theory of quantum gravity should replace our present concept of spacetime at a singularity. It should give a clear-cut way of talking about what we call a singularity in classical theory. It shouldn't be simply a nonsingular spacetime, but something drastically different.

CHAPTER THREE

Quantum Black Holes

S. W. Hawking

IN MY SECOND LECTURE, I'm going to talk about the quantum theory of black holes. It seems to lead to a new level of unpredictability in physics over and above the usual uncertainty associated with quantum mechanics. This is because black holes appear to have intrinsic entropy and to lose information from our region of the universe. I should say that these claims are controversial: many people working on quantum gravity, including almost all those who entered it from particle physics, would instinctively reject the idea that information about the quantum state of a system could be lost. However, they have had very little success in showing how information can get out of a black hole. Eventually I believe they will be forced to accept my suggestion that it is lost, just as they were forced to agree that black holes radiate, which went against all their preconceptions.

I should start by reminding you about the classical theory of black holes. We saw in the last lecture that gravity is always attractive, at least in normal situations. If gravity had been sometimes attractive and sometimes repulsive, like electrodynamics, we would never notice it at all because it is about 10^{40} times weaker. It is only because gravity always has the same sign that the gravitational force between the particles of two macroscopic bodies like ourselves and the Earth add up to give a force we can feel.

The fact that gravity is attractive means that it will tend to draw the matter in the universe together to form objects like stars and galaxies. These can support themselves for a time against further contraction by thermal pressure, in the case of stars, or by rotation and internal motions, in the case of galaxies. However, eventually the

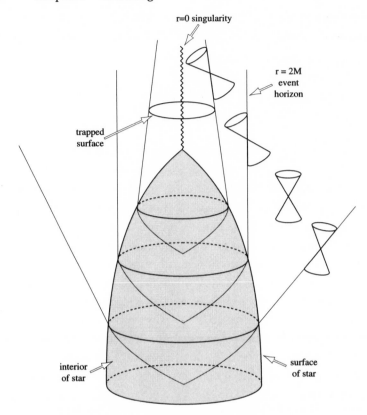

r=0 singularity

r = 2M
event
horizon

trapped
surface

interior
of star

surface
of star

Figure 3.1 A spacetime picture of the collapse of a star to form a black hole, showing the event horizon and a closed trapped surface.

heat or the angular momentum will be carried away and the object will begin to shrink. If the mass is less than about one and a half times that of the Sun, the contraction can be stopped by the degeneracy pressure of electrons or neutrons. The object will settle down to be a white dwarf or a neutron star, respectively. However, if the mass is greater than this limit there is nothing that can hold it up and stop it continuing to contract. Once it has shrunk to a certain critical size the gravitational field at its surface will be so strong that the light cones will be bent inward, as in figure 3.1. I would have liked to draw you a four-dimensional picture. However, government cuts have meant that Cambridge University can afford only two-dimensional screens. I have therefore shown time in the vertical direction and used perspective to show two of the three space directions. You can see that even the outgoing light rays are bent toward each other and so are converging rather than diverging. This means that there is a closed trapped surface, which is one of the alternative third conditions of the Hawking-Penrose theorem.

If the cosmic censorship conjecture is correct, the trapped surface and the singularity it predicts cannot be visible from far away. Thus there must be a region of spacetime from which it is not possible to escape to infinity. This region is said to be a black hole. Its boundary is called the event horizon and is a null surface formed by the light rays that just fail to get away to infinity. As we saw in the last lecture, the area of a cross section of the event horizon can never decrease, at least in the classical theory. This, and perturbation calculations of spherical collapse, suggest that black holes will settle down to a stationary state. The no-hair theorem, proved by the combined work of Israel, Carter, Robinson, and myself, shows that the only stationary black holes in the absence of matter fields are the Kerr solutions. These are characterized by two parameters, the mass M and the angular momentum J. The no-hair theorem was extended by Robinson to the case where there was an electromagnetic field. This added a third parameter Q, the electric charge (see box 3.A). The no-hair theorem has not been proved for the Yang-Mills field, but the only difference seems to be the addition of one or more integers that label a discrete family of unstable solutions. It can be shown that there are no more continuous degrees of freedom of time-independent Einstein-Yang-Mills black holes.

What the no-hair theorems show is that a large amount of information is lost when a body collapses to form a black hole. The collapsing body is described by a very large number of parameters. There are the types of matter and the multipole moments of the mass distribution. Yet the black hole that forms is completely independent of the type of matter and rapidly loses all the multipole moments except the first two: the monopole moment, which is the mass, and the dipole moment, which is the angular momentum.

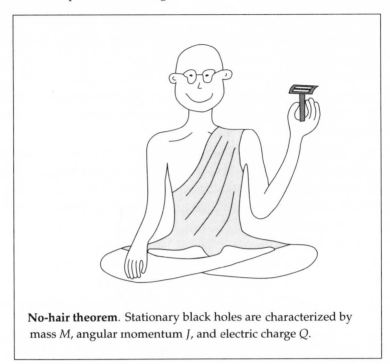

No-hair theorem. Stationary black holes are characterized by mass M, angular momentum J, and electric charge Q.

This loss of information didn't really matter in the classical theory. One could say that all the information about the collapsing body was still inside the black hole. It would be very difficult for an observer outside the black hole to determine what the collapsing body was like. However, in the classical theory it was still possible in principle. The observer would never actually lose sight of the collapsing body. Instead, it would appear to slow down and get very dim as it approached the event horizon. But the observer could still see what it was made of and how the mass was distributed. However, quantum theory changed all this. First, the collapsing body would send out only a limited number of photons before it crossed the event horizon. They would be quite insufficient to carry all the information about the collapsing body. This means that in quantum theory there's no way an outside observer can measure the state of the collapsed body. One might not think this mattered too much, because the information would still be inside the black hole even if one couldn't measure it from the outside. But this is where the second effect of quantum theory on black holes comes in. As I will show, quantum theory will cause black holes to radiate and lose mass. It seems that they will eventually disappear completely, taking with them the information inside them. I will give arguments that this information really is lost and doesn't come back in some form. As I will show, this loss of information would introduce a new level of uncertainty into physics over and above the usual uncertainty associated with quantum theory. Unfortunately, unlike Heisenberg's uncertainty principle, this extra level will be rather difficult to confirm experimentally in the case of black holes. But as I will argue in my third lecture (chapter 5), there's a sense in which we may have already observed it in the measurements of fluctuations in the microwave background.

The fact that quantum theory causes black holes to radiate was first discovered by doing quantum field theory on the background of a black hole formed by collapse. To see how this comes about it is helpful to use what are normally called Penrose diagrams. However, I think Penrose himself would agree they really should be called Carter diagrams because Carter was the first to use them systematically. In a spherical collapse the spacetime won't depend on the angles θ and ϕ. All the geometry will take place in the r-t plane. Because any two-dimensional plane is conformal to flat space one can represent the causal structure by a diagram in which null lines in the r-t plane are at $\pm 45°$ to the vertical.

Let's start with flat Minkowski space, which has a Carter-Penrose diagram that is a triangle standing on one corner (fig. 3.2). The two diagonal sides on the right correspond to the past and future null infinities I referred to in my first lecture. These are really at infinity but all distances are shrunk by a conformal factor as one approaches past or future null infinity. Each point of this triangle corresponds to a two-sphere of radius r. $r = 0$ on the vertical line on the left, which represents the center of symmetry, and $r \to \infty$ on the right of the diagram.

One can easily see from the diagram that every point in Minkowski

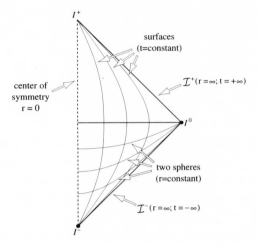

Figure 3.2 The Carter-Penrose diagram for Minkowski space.

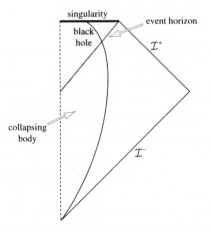

Figure 3.3 The Carter-Penrose diagram for a star that collapses to form a black hole.

space is in the past of future null infinity $\mathcal{I}^+$. This means there is no black hole and no event horizon. However, if one has a spherical body collapsing the diagram is rather different (fig. 3.3). It looks the same in the past but now the top of the triangle has been cut off and replaced by a horizontal boundary. This is the singularity that the

Hawking-Penrose theorem predicts. One can now see that there are points under this horizontal line that are not in the past of future null infinity $\mathcal{I}^+$. In other words, there is a black hole. The event horizon, the boundary of the black hole, is a diagonal line that comes down from the top right corner and meets the vertical line corresponding to the center of symmetry.

One can consider a scalar field ϕ on this background. If the space-time were time independent, a solution of the wave equation that contained only positive frequencies on $\mathcal{I}^-$ would also be of positive frequency on $\mathcal{I}^+$. This would mean that there would be no particle creation, and there would be no outgoing particles on $\mathcal{I}^+$ if there were no scalar particles initially.

However, the metric is time dependent during the collapse. This will cause a solution that is positive frequency on $\mathcal{I}^-$ to be partly negative frequency when it gets to $\mathcal{I}^+$. One can calculate this mixing by taking a wave with time dependence $e^{-i\omega u}$ on $\mathcal{I}^+$ and propagating it back to $\mathcal{I}^-$. When one does that, one finds that the part of the wave that passes near the horizon is very blue shifted. Remarkably, it turns out that the mixing is independent of the details of the collapse in the limit of late times. It depends only on the surface gravity κ that measures the strength of the gravitational field on the horizon of the black hole. The mixing of positive and negative frequencies leads to particle creation.

When I first studied this effect in 1973 I expected I would find a burst of emission during the collapse but that then the particle creation would die out and one would be left with a black hole that was truly black. To my great surprise, I found that after a burst during the collapse there remained a steady rate of particle creation and emission. Moreover, the emission was exactly thermal with a temperature of $\frac{\kappa}{2\pi}$. This was just what was required to make the idea that a black hole had an entropy proportional to the area of its event horizon consistent. Moreover, it fixed the constant of proportionality to be a quarter in Planck units, in which $G = c = \hbar = 1$. This makes the unit of area 10^{-66} cm^2, so a black hole of the mass of the Sun would have an entropy of the order of 10^{78}. This would reflect the enormous number of different ways in which it could be made.

Black Hole Thermal Radiation

$$\text{Temperature } T = \frac{\kappa}{2\pi}$$

$$\text{Entropy } S = \frac{1}{4}A$$

When I made my original discovery of radiation from black holes it seemed a miracle that a rather messy calculation should lead to emission that was exactly thermal. However, joint work with Jim Hartle and Gary Gibbons uncovered the deep reason. To explain it I shall start with the example of the Schwarzschild metric.

Schwarzschild Metric

$$ds^2 \;=\; -\left(1 - \frac{2M}{r}\right)dt^2 + \left(1 - \frac{2M}{r}\right)^{-1}dr^2$$
$$+ \, r^2(d\theta^2 + \sin^2\theta d\phi^2)$$

This represents the gravitational field that a black hole would settle down to if it were nonrotating. In the usual r and t coordinates there is an apparent singularity at the Schwarzschild radius $r = 2M$. However, this is just caused by a bad choice of coordinates. One can choose other coordinates in which the metric is regular there.

The Carter-Penrose diagram has the form of a diamond with flattened top and bottom (fig. 3.4). It is divided into four regions by the two null surfaces on which $r = 2M$. The region on the right, marked ① on the diagram, is the asymptotically flat space in which we are supposed to live. It has past and future null infinities $\mathcal{I}^-$ and $\mathcal{I}^+$ like flat spacetime. There is another asymptotically flat region ③ on the

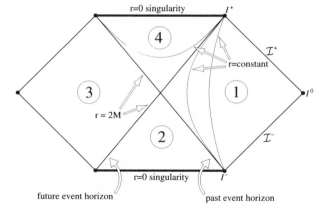

Figure 3.4 The Carter-Penrose diagram of an eternal Schwarzschild black hole.

left that seems to correspond to another universe that is connected to ours only through a wormhole. However, as we shall see, it is connected to our region through imaginary time. The null surface from bottom left to top right is the boundary of the region from which one can escape to the infinity on the right. Thus, it is the future event horizon, the epithet future being added to distinguish it from the past event horizon which goes from bottom right to top left.

Let us return to the Schwarzschild metric in the original r and t coordinates. If one puts $t = i\tau$ one gets a positive definite metric. I shall refer to such positive definite metrics as Euclidean, even though they may be curved. In the Euclidean-Schwarzschild metric there is again an apparent singularity at $r = 2M$. However, one can define a new radial coordinate x to be $4M(1 - 2Mr^{-1})^{\frac{1}{2}}$.

Euclidean-Schwarzschild Metric

$$ds^2 = x^2 \left(\frac{d\tau}{4M} \right)^2 + \left(\frac{r^2}{4M^2} \right)^2 dx^2 + r^2(d\theta^2 + \sin^2\theta \, d\phi^2)$$

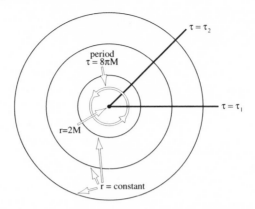

Figure 3.5 The Euclidean-Schwarzschild solution, in which τ is identified periodically.

The metric in the $x-\tau$ plane then becomes like the origin of polar coordinates if one identifies the coordinate τ with period $8\pi M$. Similarly, other Euclidean black hole metrics will have apparent singularities on their horizons that can be removed by identifying the imaginary time coordinate with period $\frac{2\pi}{\kappa}$ (fig. 3.5).

So what is the significance of having imaginary time identified with some period β? To see this, consider the amplitude to go from some field configuration ϕ_1 on the surface t_1 to a configuration ϕ_2 on the surface t_2. This will be given by the matrix element of $e^{-iH(t_2-t_1)}$. However, one can also represent this amplitude as a path integral over all fields ϕ between t_1 and t_2 that agree with the given fields ϕ_1 and ϕ_2 on the two surfaces (fig. 3.6).

One now chooses the time separation $(t_2 - t_1)$ to be pure imaginary and equal to β (fig. 3.7). One also puts the initial field ϕ_1 equal to the final field ϕ_2 and sums over a complete basis of states ϕ_n. On the left one has the expectation value of $e^{-\beta H}$ summed over all states. This is just the thermodynamic partition function Z at the temperature $T = \beta^{-1}$.

On the right hand of the equation one has a path integral. One puts $\phi_1 = \phi_2$ and sums over all field configurations ϕ_n. This means that

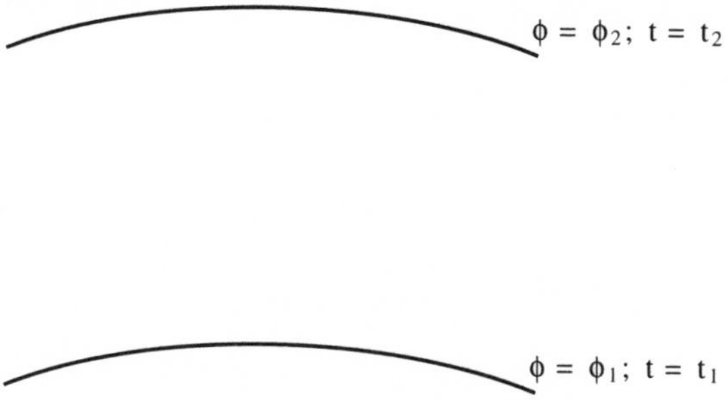

$$\phi = \phi_2; \; t = t_2$$

$$\phi = \phi_1; \; t = t_1$$

$$\langle \phi_2, t_2 \mid \phi_1, t_1 \rangle \quad = \quad \langle \phi_2 \mid \exp(-iH(t_2 - t_1)) \mid \phi_1 \rangle$$

$$= \quad \int D[\phi] \exp(iI[\phi])$$

Figure 3.6 The amplitude to go from the state ϕ_1 at t_1 to ϕ_2 at t_2.

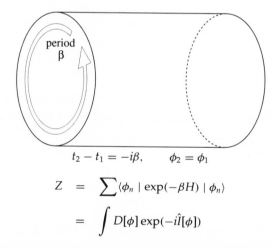

$$t_2 - t_1 = -i\beta, \qquad \phi_2 = \phi_1$$

$$Z \quad = \quad \sum \langle \phi_n \mid \exp(-\beta H) \mid \phi_n \rangle$$

$$= \quad \int D[\phi] \exp(-i\hat{I}[\phi])$$

Figure 3.7 The partition function at temperature T is given by the path integral over all fields on a Euclidean spacetime with period $\beta = T^{-1}$ in the imaginary time direction.

effectively one is doing the path integral over all fields ϕ on a space-time that is identified periodically in the imaginary time direction with period β. Thus the partition function for the field ϕ at temperature T is given by a path integral over all fields on a Euclidean spacetime. This spacetime is periodic in the imaginary time direction with period $\beta = T^{-1}$.

If one does the path integral in flat spacetime identified with period β in the imaginary time direction, one gets the usual result for the partition function of black body radiation. However, as we have just seen, the Euclidean-Schwarzschild solution is also periodic in imaginary time with period $\frac{2\pi}{\kappa}$. This means that fields on the Schwarzschild background will behave as if they were in a thermal state with temperature $\frac{\kappa}{2\pi}$.

The periodicity in imaginary time explained why the messy calculation of frequency mixing led to radiation that was exactly thermal. However, this derivation avoided the problem of the very high frequencies that take part in the frequency mixing approach. It can also be applied when there are interactions between the quantum fields on the background. The fact that the path integral is on a periodic background implies that all physical quantities such as expectation values will be thermal. This would have been very difficult to establish in the frequency mixing approach.

One can extend these interactions to include interactions with the gravitational field itself. One starts with a background metric g_0 such as the Euclidean-Schwarzschild metric that is a solution of the classical field equations. One can then expand the action I in a power series in the perturbations δg about g_0:

$$I[g] = I[g_0] + I_2(\delta g)^2 + I_3(\delta g)^3 + \cdots$$

The linear term vanishes because the background is a solution of the field equations. The quadratic term can be regarded as describing gravitons on the background, while the cubic and higher terms describe interactions between the gravitons. The path integral over the quadratic terms is finite. There are nonrenormalizable divergences at two loops in pure gravity, but these cancel with the fermions in supergravity theories. It is not known whether supergravity theories have divergences at three loops or higher because no one has been brave or foolhardy enough to try the calculation. Some recent work indicates that they may be finite to all orders. But even if there are higher loop divergences they will make very little difference except when the background is curved on the scale of the Planck length, 10^{-33} cm.

More interesting than the higher-order terms is the zeroth order term, the action of the background metric g_0:

$$I = -\frac{1}{16\pi} \int R(-g)^{\frac{1}{2}} d^4x + \frac{1}{8\pi} \int K(\pm h)^{\frac{1}{2}} d^3x.$$

The usual Einstein-Hilbert action for general relativity is the volume integral of the scalar curvature R. This is zero for vacuum solutions, so one might think that the action of the Euclidean-Schwarzschild solution was zero. However, there is also a surface term in the action proportional to the integral of K, the trace of the second fundamental form of the boundary surface. When one includes this and subtracts off the surface term for flat space, one finds that the action of the Euclidean-Schwarzschild metric is $\frac{\beta^2}{16\pi}$, where β is the period in imaginary time at infinity. Thus the dominant contribution to the path integral for the partition function Z is $e^{\frac{-\beta^2}{16\pi}}$:

$$Z = \sum \exp(-\beta E_n) = \exp\left(-\frac{\beta^2}{16\pi}\right).$$

If one differentiates $\log Z$ with respect to the period β, one gets the expectation value of the energy or, in other words, the mass:

$$\langle E \rangle = -\frac{d}{d\beta}(\log Z) = \frac{\beta}{8\pi}.$$

So this gives the mass $M = \frac{\beta}{8\pi}$. This confirms the relation between the mass and the period, or inverse temperature, that we already knew. However, one can go further. By standard thermodynamic arguments, the log of the partition function is equal to minus the free energy F divided by the temperature T:

$$\log Z = -\frac{F}{T}.$$

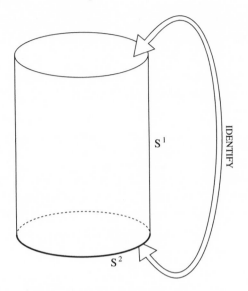

Figure 3.8 The boundary at infinity in the Euclidean-Schwarzschild solution.

And the free energy is the mass or energy plus the temperature times the entropy S:

$$F = \langle E \rangle + TS.$$

Putting all this together one sees that the action of the black hole gives an entropy of $4\pi M^2$:

$$S = \frac{\beta^2}{16\pi} = 4\pi M^2 = \frac{1}{4}A.$$

This is exactly what is required to make the laws of black holes the same as the laws of thermodynamics.

Why does one get this intrinsic gravitational entropy which has no parallel in other quantum field theories? The reason is gravity allows different topologies for the spacetime manifold. In the case we are considering, the Euclidean-Schwarzschild solution has a boundary at infinity that has topology $S^2 \times S^1$. The S^2 is a large spacelike two-sphere at infinity and the S^1 corresponds to the imaginary time

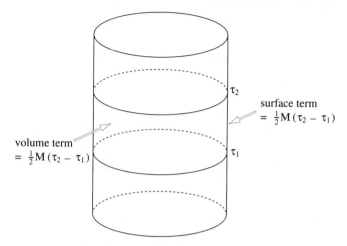

Figure 3.9 The action of periodically identified Euclidean flat space $= M(\tau_2 - \tau_1)$.

direction which is identified periodically (fig. 3.8). One can fill in this boundary with metrics of at least two different topologies. One of course is the Euclidean-Schwarzschild metric. This has topology $R^2 \times S^2$, that is, the Euclidean two-plane times a two-sphere. The other is $R^3 \times S^1$, the topology of Euclidean flat space periodically identified in the imaginary time direction. These two topologies have different Euler numbers. The Euler number of periodically identified flat space is zero, while that of the Euclidean-Schwarzschild solution is two. The significance of this is as follows: on the topology of periodically identified flat space, one can find a periodic time function τ whose gradient is nowhere zero and which agrees with the imaginary time coordinate on the boundary at infinity. One can then work out the action of the region between two surfaces τ_1 and τ_2. There will be two contributions to the action, a volume integral over the matter Lagrangian plus the Einstein-Hilbert Lagrangian and a surface term. If the solution is time independent, the surface term over $\tau = \tau_1$ will cancel with the surface term over $\tau = \tau_2$. Thus the only net contribution to the surface term comes from the boundary at infinity. This gives half the mass times the imaginary time interval

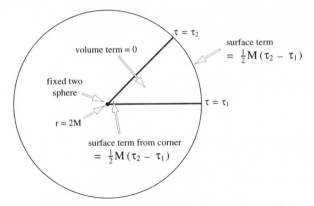

$$\text{Total action including corner contribution} = M(\tau_2 - \tau_1)$$
$$\text{Total action without corner contribution} = \frac{1}{2}M(\tau_2 - \tau_1)$$

Figure 3.10 The total action for the Euclidean-Schwarzschild solution $= \frac{1}{2}M(\tau_2 - \tau_1)$, as we don't include the corner contribution from $r = 2M$.

$(\tau_2 - \tau_1)$. If the mass is nonzero there must be nonzero matter fields to create the mass. One can show that the volume integral over the matter Lagrangian plus the Einstein-Hilbert Lagrangian also gives $\frac{1}{2}M(\tau_2 - \tau_1)$. Thus the total action is $M(\tau_2 - \tau_1)$ (fig. 3.9). If one puts this contribution to the log of the partition function into the thermodynamic formulae, one finds the expectation value of the energy to be the mass, as one would expect. However, the entropy contributed by the background field will be zero.

The situation is different, however, with the Euclidean-Schwarzschild solution. Because the Euler number is two rather than zero, one can't find a time function τ whose gradient is everywhere nonzero. The best one can do is choose the imaginary time coordinate of the Schwarzschild solution. This has a fixed two-sphere at the horizon, where τ behaves like an angular coordinate. If one now works out the action between two surfaces of constant τ, the volume integral vanishes because there are no matter fields and the scalar curvature is zero. The trace K surface term at infinity again gives $\frac{1}{2}M(\tau_2 - \tau_1)$. However, there is now another surface term at the horizon where the

τ_1 and τ_2 surfaces meet in a corner. One can evaluate this surface term and find that it also is equal to $\frac{1}{2}M(\tau_2 - \tau_1)$ (fig. 3.10). Thus the total action for the region between τ_1 and τ_2 is $M(\tau_2 - \tau_1)$. If one used this action with $\tau_2 - \tau_1 = \beta$ one would find that the entropy was zero. However, when one looks at the action of the Euclidean-Schwarz-schild solution from a four-dimensional point of view rather than a $3 + 1$, there is no reason to include a surface term on the horizon, because the metric is regular there. Leaving out the surface term on the horizon reduces the action by one quarter the area of the horizon, which is just the intrinsic gravitational entropy of the black hole.

The fact that the entropy of black holes is connected with a topological invariant, the Euler number, is a strong argument that it will remain even if we have to go to a more fundamental theory. This idea is anathema to most particle physicists, who are a very conservative lot and want to make everything like the Yang-Mills theory. They agree that the radiation from black holes seems to be thermal and independent of how the hole was formed if the hole is large compared to the Planck length. But they would claim that when the black hole loses mass and gets down to the Planck size, quantum general relativity will break down and all bets will be off. However, I shall describe a thought experiment with black holes in which information seems to be lost, yet the curvature outside the horizons always remains small.

It has been known for some time that one can create pairs of positively and negatively charged particles in a strong electric field. One way of looking at this is to note that in flat Euclidean space a particle of charge q such as an electron would move in a circle in a uniform electric field E. One can analytically continue this motion from the imaginary time τ to real time t. One gets a pair of positively and negatively charged particles accelerating away from each other pulled apart by the electric field (fig. 3.11).

The process of pair creation is described by chopping the two diagrams in half along the $t = 0$ or $\tau = 0$ lines. One then joins the upper half of the Minkowski space diagram to the lower half of the Euclidean space diagram (fig. 3.12). This gives a picture in which the positively and negatively charged particles are really the same particle. It tunnels through Euclidean space to get from one Minkowski

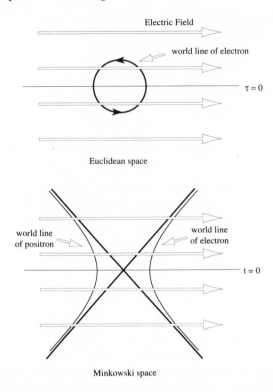

Figure 3.11 In Euclidean space, an electron moves on a circle in an electric field. In Minkowski space, we get a pair of oppositely charged particles accelerating away from each other.

space world line to the other. To a first approximation the probability for pair creation is e^{-I}, where

$$\text{Euclidean action } I = \frac{2\pi m^2}{qE}.$$

Pair creation by strong electric fields has been observed experimentally and the rate agrees with these estimates.

Black holes can also carry electric charges so one might expect that they could also be pair created. However, the rate would be tiny compared to that for electron-positron pairs because the mass-to-charge

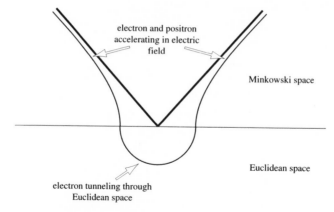

electron and positron
accelerating in electric
field

Minkowski space

Euclidean space

electron tunneling through
Euclidean space

Figure 3.12 Pair creation is described by joining half the Euclidean diagram to half the Minkowski diagram.

ratio is 10^{20} times bigger. This means that any electric field would be neutralized by electron-positron pair creation long before there was a significant probability of pair creating black holes. However, there are also black hole solutions with magnetic charges. Such black holes couldn't be produced by gravitational collapse because there are no magnetically charged elementary particles. But one might expect that they could be pair created in a strong magnetic field. In this case there would be no competition from ordinary particle creation because ordinary particles do not carry magnetic charges. So the magnetic field could become strong enough that there was a significant chance of creating a pair of magnetically charged black holes.

In 1976 Ernst found a solution that represented two magnetically charged black holes accelerating away from each other in a magnetic field (fig. 3.13). If one analytically continues it to imaginary time, one has a picture very like that of the electron pair creation (fig. 3.14). The black hole moves on a circle in a curved Euclidean space just like the electron moves in a circle in flat Euclidean space. There is a complication in the black hole case, because the imaginary time coordinate is periodic about the horizon of the black hole as well as about the center of the circle on which the black hole moves. One

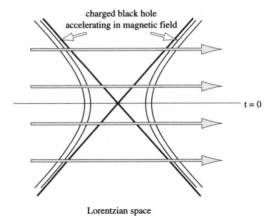

Lorentzian space

Figure 3.13 A pair of oppositely charged black holes accelerating away from each other in a magnetic field.

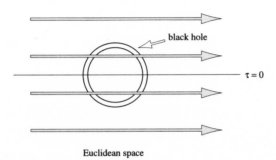

Euclidean space

Figure 3.14 A charged black hole moving on a circle in Euclidean space.

has to adjust the mass-to-charge ratio of the black hole to make these periods equal. Physically this means that one chooses the parameters of the black hole so that the temperature of the black hole is equal to the temperature it sees because it is accelerating. The temperature of a magnetically charged black hole tends to zero as the charge tends to the mass in Planck units. Thus for weak magnetic fields, and hence low acceleration, one can always match the periods.

As in the case of pair creation of electrons, one can describe pair creation of black holes by joining the lower half of the imaginary time Euclidean solution to the upper half of the real time Lorentzian solu-

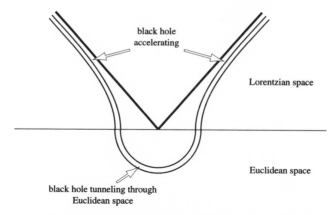

Figure 3.15 Tunneling to produce a pair of black holes is also described by joining half the Euclidean diagram to half the Lorentzian diagram.

tion (fig. 3.15). One can think of the black hole as tunneling through the Euclidean region and emerging as a pair of oppositely charged black holes that accelerate away from each other pulled apart by the magnetic field. The accelerating black hole solution is not asymptotically flat because it tends to a uniform magnetic field at infinity. But one can nevertheless use it to estimate the rate of pair creation of black holes in a local region of magnetic field. One could imagine that after being created, the black holes move far apart into regions without magnetic field. One could then treat each black hole separately as a black hole in asymptotically flat space. One could throw an arbitrarily large amount of matter and information into each hole. The holes would then radiate and lose mass. However, they couldn't lose magnetic charge because there are no magnetically charged particles. Thus they would eventually get back to their original state with the mass slightly bigger than the charge. One could then bring the two holes back together again and let them annihilate each other. The annihilation process can be regarded as the time reverse of the pair creation. Thus it is represented by the top half of the Euclidean solution joined to the bottom half of the Lorentzian solution. In between the pair creation and the annihilation, one can have a long

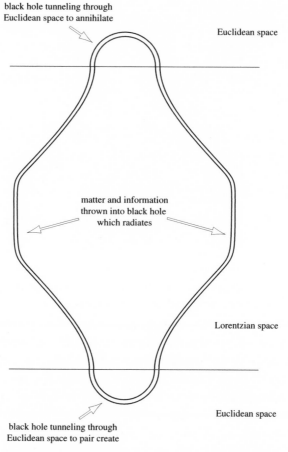

black hole tunneling through
Euclidean space to annihilate

Euclidean space

matter and information
thrown into black hole
which radiates

Lorentzian space

Euclidean space

black hole tunneling through
Euclidean space to pair create

Figure 3.16 A pair of black holes are produced by tunneling and eventually annihilated again by tunneling.

Lorentzian period in which the black holes move far apart, accrete matter, radiate, and then come back together again. But the topology of the gravitational field will be the topology of the Euclidean-Ernst solution. This is $S^2 \times S^2$ minus a point (fig. 3.16).

One might worry that the generalized second law of thermodynamics would be violated when the black holes annihilated because the black hole horizon area would have disappeared. However, it turns out that the area of the acceleration horizon in the Ernst solution is reduced from the area it would have if there were no pair creation. This is a rather delicate calculation because the area of the acceleration horizon is infinite in both cases. Nevertheless, there is a well-defined sense in which their difference is finite and equal to the black hole horizon area plus the difference in the action of the solutions with and without pair creation. This can be understood as saying that pair creation is a zero energy process; the Hamiltonian *with* pair creation is the same as the Hamiltonian *without*. I'm very grateful to Simon Ross and Gary Horowitz for calculating this reduction just in time for this lecture. It is miracles like this—and I mean the result, not that they got it—that convince me that black hole thermodynamics can't just be a low energy approximation. I believe that gravitational entropy won't disappear even if we have to go to a more fundamental theory of quantum gravity.

One can see from this thought experiment that one gets intrinsic gravitational entropy and loss of information when the topology of spacetime is different from that of flat Minkowski space. If the black holes that are pair created are large compared to the Planck size, the curvature outside the horizons will be everywhere small compared to the Planck scale. This means that the approximation I have made of ignoring cubic and higher terms in the perturbations should be good. Thus the conclusion that information can be lost in black holes should be reliable.

If information is lost in macroscopic black holes it should also be lost in processes in which microscopic, virtual black holes appear because of quantum fluctuations of the metric. One could imagine that particles and information could fall into these holes and get lost. Maybe that is where all those odd socks went. Quantities like energy and electric charge that are coupled to gauge fields would be conserved, but other information and global charge would be lost. This would have far-reaching implications for quantum theory.

It is normally assumed that a system in a pure quantum state evolves in a unitary way through a succession of pure quantum states. But if there is loss of information through the appearance and disappearance of black holes, there can't be a unitary evolution. Instead, the loss of information will mean that the final state after the black

Figure 3.17

holes have disappeared will be what is called a *mixed quantum state*. This can be regarded as an ensemble of different pure quantum states, each with its own probability. But because it is not with certainty in any one state, one cannot reduce the probability of the final state to zero by interfering with any quantum state. This means that gravity introduces a new level of unpredictability into physics over and above the uncertainty usually associated with quantum theory. I shall show in the next lecture (chapter 5) that we may have already observed this extra uncertainty. It means an end to the hope of scientific determinism, that we could predict the future with certainty. It seems God still has a few tricks up his sleeve (fig. 3.17).

CHAPTER FOUR

Quantum Theory and Spacetime

R. Penrose

THE GREAT PHYSICAL THEORIES of the twentieth century have been quantum theory (QT), special relativity (SR), general relativity (GR), and quantum field theory (QFT). These theories are not independent of each other: general relativity was built on special relativity, and quantum field theory has special relativity and quantum theory as inputs (see fig. 4.1).

It has been said that quantum field theory is the most accurate physical theory ever, being accurate to about one part in about 10^{11}. However, I would like to point out that general relativity has, in a certain clear sense, now been tested to be correct to one part in 10^{14} (and this accuracy has apparently been limited merely by the accuracy of clocks on earth). I am speaking of the Hulse-Taylor binary pulsar PSR 1913 + 16, a pair of neutron stars orbiting each other, one of which is a pulsar. GR predicts that this orbit will slowly decay (and the period shorten) because energy is lost through the emission of gravitational waves. This has indeed been observed, and the entire description of the motion, incorporating the Newtonian orbits at one end of the scale, with GR corrections at the middle range, up to the orbital speedup due to gravitational radiation at the other, agrees with GR (which I am taking to include Newtonian theory) to the remarkable accuracy, noted above, over an accumulated period of twenty years. The discoverers of this system have now rightly been awarded Nobel prizes for their work. The quantum theorists have always claimed that because of the accuracy of their theory, it should be GR that is changed to fit their mold, but I think now that it is QFT that has some catching up to do.

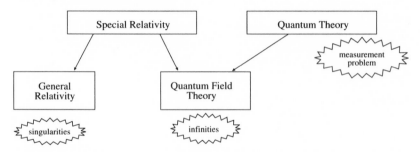

Figure 4.1 The great physical theories of the twentieth century—and their fundamental problems.

Although these four theories have been remarkably successful, they are not without their problems. QFT has the problem that whenever you calculate the amplitude for a multiply-connected Feynman diagram, the answer is infinity. These infinities must be subtracted away or scaled away as part of the process of renormalization of the theory. GR predicts the existence of spacetime singularities. In QT there is the "measurement problem"—I shall describe this later. It may be taken that the solution to the various problems of these theories lies in the fact that they are incomplete on their own. For example, it is anticipated by many that QFT might "smear" out the singularities of GR in some way. The divergence problems in QFT could be solved in part by an ultraviolet cutoff from GR. I believe that the measurement problem, likewise, will ultimately be resolved when GR and QT are appropriately combined in some new theory.

I should now like to talk about information loss in black holes, which I claim is relevant to this last issue. I agree with nearly all that Stephen had to say on this. But while Stephen regards the information loss due to black holes as an extra uncertainty in physics, above and beyond the uncertainty from QT, I regard it as a "complementary" uncertainty. Let me explain what I mean by this. In a spacetime with a black hole, one may see how the information loss happens by constructing a Carter diagram of the spacetime (fig. 4.2). The "in information" is specified on past null infinity $\mathcal{I}^-$ and, the "out information" on future null infinity $\mathcal{I}^+$. One could say that the missing information is lost when it falls through the horizon of the black hole,

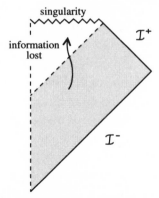

Figure 4.2 Carter diagram of collapse of black hole.

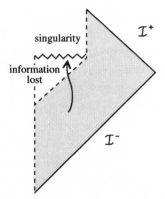

Figure 4.3 Carter diagram of evaporating black hole.

but I prefer to regard it as lost when it meets the singularity. Now consider a collapse of a body of matter to a black hole, followed by the evaporation of the black hole by Hawking radiation. (One would certainly have to wait a long time for this to happen—maybe longer than the lifetime of the universe!) I agree with Stephen's view that information is lost in this collapse and evaporation picture. We can also draw a Carter diagram of this entire spacetime (fig. 4.3).

The singularity inside the black hole is spacelike and has a large Weyl curvature, in accordance with my previous lecture's discussion (chapter 2). It is possible that a little bit of information escapes at the moment of the black hole evaporation, from a residual piece of singularity (which, as it will be to the past of future external observers, will have little or no Weyl curvature), but this tiny information gain will be much smaller than the information loss in the collapse (in what I regard as any reasonable picture of the hole's final disappearance). If we enclose this system in a vast box, as a thought experiment, we can consider the phase-space evolution of matter inside the box. In the region of phase space corresponding to situations in which a black hole is present, trajectories of physical evolution will converge and volumes following these trajectories will shrink. This is due to the information lost into the singularity in the black hole. This shrinking is in direct contradiction to the theorem in ordinary classical mechanics, called *Liouville's theorem*, which says that volumes in phase space remain constant. (This is a classical theorem. Strictly speaking, we should be considering a quantum evolution in Hilbert space. The violation of Liouville's theorem would then correspond to a nonunitary evolution.) Thus a black hole spacetime violates this conservation. However, in my picture, this loss of phase-space volume is balanced by a process of "spontaneous" quantum measurement in which information is gained and phase-space volumes increase. This is why I regard the uncertainty due to information loss in black holes as being "complementary" to the uncertainty in quantum theory: one is the other side of the coin to the other (see fig. 4.4).

One may say that past singularities carry little information whereas future singularities carry lots. This is what underlies the second law of thermodynamics. The asymmetry in these singularities is also related to the asymmetry of the measurement process. So let us next return to the problem of measurement in quantum theory.

The two-slit problem can be used to illustrate the principles of quantum theory. In this situation a beam of light is shone at an opaque barrier with two slits A and B in it. This produces an interference pattern of bright and dark bands on a screen behind. Individual photons reach the screen at discrete points, but because of the interference bands there are points on the screen that cannot be reached. Let p be such a point—nevertheless, p *could* be reached if one or the other of the slits were blocked off. Destructive interference of this na-

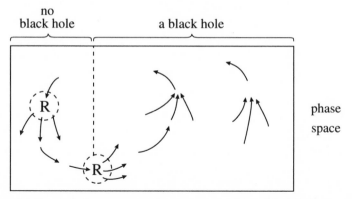

Figure 4.4 Loss of phase-space volume occurs when a black hole is present. This may be balanced against regain of phase-space volume due to wave function collapse **R**.

ture, where alternative possibilities can sometimes cancel out, is one of the most puzzling features of quantum mechanics. We understand this in terms of quantum theory's *superposition principle*, which says that if route A and route B are possible for the photon, with respective photon states denoted by $|A\rangle$ and $|B\rangle$—and let us suppose that these are routes that the photon might take to reach p, by first passing through one slit or first passing through the other—then a combination $z|A\rangle + w|B\rangle$ is also possible where z and w are complex numbers.

It is inappropriate to regard w and z as being in any way *probabilities* since they are *complex numbers*. The state of the photon *is* just such a complex superposition. *Unitary* evolution of a quantum system (which I call **U**) preserves the superpositions: if $zA_0 + wB_0$ is a superposition at time $t = 0$, then after a time t this will have evolved to $zA_t + wB_t$, where A_t and B_t represent the separate evolutions of the two alternatives after time t. Upon measurement of a quantum system, where quantum alternatives are magnified to give distinguishable classical outcomes, a different kind of "evolution" appears to take place, called *reduction* of the state vector or "collapse of wave function" (I shall call this **R**). Probabilities only enter when the system is "measured," in this sense and the relative likelihoods for the two events to occur is $|z|^2 : |w|^2$.

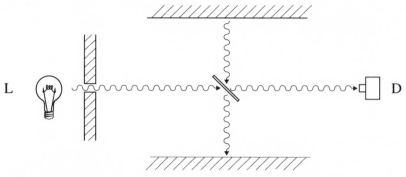

Figure 4.5 A simple experiment that illustrates that the quantum probabilities inherent in **R** do not apply in the reverse time direction.

U and **R** are very different processes: **U** is deterministic, linear, local (in configuration space), and time symmetric. **R** is nondeterministic, decidedly nonlinear, nonlocal, and time asymmetric. This difference between the two fundamental evolution processes in QT is remarkable. It is most unlikely that **R** might ever be deduced as an approximation to **U** (although people often try to do this). This is the "measurement problem."

R is, in particular, time asymmetric. Suppose a beam of light from a photon source L is shone at a half silvered mirror, angled at 45° downward, with a detector *D* behind the mirror (fig. 4.5).

Because the mirror is only half-silvered, there is an equally weighted superposition of transmitted and reflected states. This leads to a 50% probability that any individual photon will activate the detector rather than being absorbed by the laboratory floor. This 50% is the answer to the question: "If L emits a photon, what is the probability that D receives it?" The answer to this kind of question is determined by the rule **R**. However, we could also ask "If D receives a photon, what is the probability that it was emitted by L?" One might think that we could work out probabilities in the same way as previously. **U** is time symmetric, so should this not also apply to **R**? However, applied to the past, the (time-reversed) rule **R** does not give the right probabilities. In fact, the answer to this question is determined by quite a different consideration, namely, the second law of thermodynamics—here applied to the wall—and the asymmetry is due ultimately to the asymmetry of the universe in time. Aharonov, Bergmann, and Liebowitz (1964) have shown how to fit the measurement process into a time-symmetric framework. According to this scheme, the time asymmetry of **R** would arise from the asymmetric boundary conditions in the future and past. This general framework is also the one adopted by Griffiths (1984), Omnés (1992), and Gell Mann and Hartle (1990). Since the origin of the second law can be traced back to the asymmetry in spacetime-singularity structure, this relationship suggests that the measurement problem of QT and the singularity problem of GR are related. Recall that I proposed in the last lecture that the initial singularity has very little information and vanishing Weyl tensor, whereas the final singularity (or singularities, or infinity) carries lots of information and has diverging Weyl tensor (in the case of singularities).

In order to make my own position clearer with regard to the relationship between QT and GR, I should now like to discuss what we mean by *quantum reality*: Is it true that the state vector is "real," or is the density matrix "real"? The density matrix represents our incomplete knowledge of the state and thus contains two types of probabilities—classical uncertainty as well as quantum probability. We may write the density matrix as

$$D = \sum_{i=1}^{N} p_i |\psi_i\rangle \langle \psi_i|,$$

where the p_i are probabilities, real numbers subject to $\sum p_i = 1$, and each $|\psi_i\rangle$ is normalized to unity. This is a weighted probability mixture of states. Here the $|\psi_i\rangle$ need not be orthogonal, and N may be larger than the dimension of the Hilbert space. As an example, let us consider an EPR-type experiment where a particle of spin zero, at rest in the center of the experiment, decays into two particles of spin half. These two particles fly off in opposite directions and are detected "here" and "there"—where "there" may be a long way from "here," say on the moon. We write the state vector as a superposition of possibilities:

$$|\psi\rangle = \{|\text{up here}\rangle|\text{down there}\rangle - |\text{down here}\rangle|\text{up there}\rangle\}/\sqrt{2}, \quad (4.1)$$

where |up here⟩ is a state with the spin of the particle "here" pointing in the "up" direction, and so on. Suppose now that the z-direction of the spin has been measured on the moon without our knowing about the result. Then the state here is described by the density matrix

$$D = \frac{1}{2}|\text{up here}\rangle\langle\text{up here}| + \frac{1}{2}|\text{down here}\rangle\langle\text{down here}|. \quad (4.2)$$

Alternatively, the x-direction of the spin might have been measured on the moon. Rewriting the state vector (4.1) as

$$|\psi\rangle = \{|\text{left here}\rangle|\text{right there}\rangle - |\text{right here}\rangle|\text{left there}\rangle\}/\sqrt{2},$$

we obtain the density matrix that is now appropriate

$$D = \frac{1}{2}|\text{left here}\rangle\langle\text{left here}| + \frac{1}{2}|\text{right here}\rangle\langle\text{right here}|,$$

which is in fact equal to (4.2). However, if the state vector describes reality, then the density matrix doesn't say what's going on. It just gives the results of measurement "here" provided you don't know what's going on "there." In particular, it might be possible that I get a letter from the moon informing me about the nature and result of the measurement there. Thus, if I can (in principle) obtain this information, then I do have to describe the entire (entangled) system by a state vector.

In general, there are lots of different ways of writing a given density matrix as a probability mixture of states. Moreover, by a recent theorem due to Hughston, Jozsa, and Wooters (1993), for any density matrix whatever, arising in this way as the "here" past of an EPR system, and for any interpretation of this density matrix as a probability mixture of states, there always exists a measurement "there" which gives rise precisely to this *particular* interpretation of the density matrix "here" as a probability mixture.

On the other hand, one might argue that the density matrix describes reality which, as I understand it, is closer to Stephen's view, when a black hole is present.

John Bell sometimes referred to the standard description of the process of reduction of the state vector as FAPP, which is an acronym for "for all practical purposes." According to this standard procedure, we may write the total state vector as

$$|\psi_{tot}\rangle = w|\text{up here}\rangle|?\rangle + z|\text{down here}\rangle|?'\rangle,$$

where $|?\rangle$s describe things in the environment, outside our measurement. If information is lost in the environment, then the density matrix is the best we can do:

$$D = |w|^2|\text{up here}\rangle\langle\text{up here}| + |z|^2|\text{down here}\rangle\langle\text{down here}|.$$

So long as the information from the environment cannot be retrieved, we "might as well" (FAPP) consider the state as $|\text{up here}\rangle$ or $|\text{down here}\rangle$, with probabilities $|w|^2$ and $|z|^2$, respectively.

However, we still need another assumption, as the density matrix doesn't tell us which states it is made of. To explain this point let us consider the Schrödinger's cat thought experiment. It describes the plight of a cat in a box, where (let us say) a photon is emitted which encounters a half-silvered mirror, and the transmitted part of the photon's wavefunction encounters a detector which, if it detects the photon, automatically fires a gun killing the cat. If it fails to detect the photon, then the cat is alive and well. (I know Stephen does not approve of mistreating cats, even in a thought experiment!) The wave function of the system is a superposition of these two possibilities:

$$w|\text{dead cat}\rangle|\text{bang}\rangle + z|\text{live cat}\rangle|\text{no bang}\rangle,$$

where $|\text{bang}\rangle$ and $|\text{no bang}\rangle$ refer to the environment states.

In the many-worlds view of quantum mechanics this would be (ignoring the environment)

$$w|\text{dead cat}\rangle|\text{know cat is dead}\rangle + z|\text{live cat}\rangle|\text{know cat is alive}\rangle, \quad (4.3)$$

where the $|\text{know} \cdots\rangle$ states refer to the experimenter's state of mind. But why does our perception not allow us to perceive macroscopic *superpositions*, of states such as these, and not just the macroscopic *alternatives* "cat is dead" and "cat is alive"? For example, in the case $w = z = 1/\sqrt{2}$, we can rewrite the state (4.3) as the superposition

$$\begin{aligned}
\{&(|\text{dead cat}\rangle + |\text{live cat}\rangle) \\
&\times (|\text{know cat is dead}\rangle + (|\text{know cat is alive}\rangle) \\
+ &(|\text{dead cat}\rangle - |\text{live cat}\rangle) \\
&\times (|\text{know cat is dead}\rangle - |\text{know cat is alive}\rangle)\}/2\sqrt{2}
\end{aligned}$$

so, unless we have reason to exclude "perception states" such as $(|\text{know cat is dead}\rangle + |\text{know cat is alive}\rangle)/\sqrt{2}$, we are no closer to solution than before.

The same kind of thing applies to the environment, and (again in the case $w = z = 1/\sqrt{2}$, for example) we can rewrite the density matrix as the superposition

$$\begin{aligned}
D \;=\; &\frac{1}{4}(|\text{dead cat}\rangle + |\text{live cat}\rangle)(\langle\text{dead cat}| + \langle\text{live cat}|) \\
+ &\frac{1}{4}(|\text{dead cat}\rangle - |\text{live cat}\rangle)(\langle\text{dead cat}| - \langle\text{live cat}|),
\end{aligned}$$

which tells us that the "decoherence by environment" viewpoint does not explain why the cat is simply alive or dead either.

I do not want to go further into a discussion of issues of consciousness or decoherence here. In my opinion, the answer to the measurement problem lies elsewhere. I am suggesting that something goes wrong with superpositions of the alternative spacetime geometries that would occur when GR begins to become involved. Perhaps a

(i)

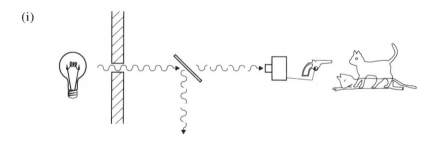

(ii)

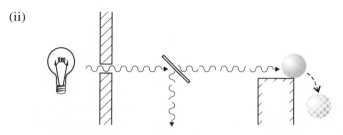

Figure 4.6 Schrödinger's cat (i), and a more humane version (ii).

superposition of two different geometries is *unstable*, and decays into *one* of the two alternatives. For example, the geometries might be the spacetimes of a live cat, or a dead one. I call this decay into one **or** the other alternative objective reduction, which I like as a name because it has an appropriately nice acronym (**OR**). How does the Planck length 10^{-33} cm relate to this? Nature's criterion for determining when two geometries are significantly different would depend upon the Planck scale, and this fixes the timescale in which the reduction into different alternatives occurs.

We may give the cat a day off, and think of the half-silvered mirror problem again, but this time with detection of a photon triggering the movement of a large piece of mass from one place to another (fig. 4.6).

We could avoid the problem of worrying about state reduction in the detector if we simply have the mass delicately poised on a cliff edge so the photon pushes it off a cliff! When is enough mass moved so that the superposition of the two alternatives becomes unstable?

Gravity may provide the answer to this, as I am indeed proposing here (cf. Penrose 1993, 1994; also Diósi 1989, Ghirardi, Grassi, and Rimini 1990). To compute the decay time, according to this proposed scheme, consider the energy E that it would cost to pull away one instance of the mass, moving it out away from coincidence, in the gravitational field of the other, until these two mass locations provide the mass superposition under consideration. I propose that the timescale of the collapse of the state vector of this superposition is of the order of

$$T \sim \frac{\hbar}{E}. \tag{4.4}$$

For a nucleon, this would be nearly 10^8 years, so the instability would not be seen in existing experiments. However, for a speck of water of 10^{-5} cm in size, the collapse would take around 2 hours. If the speck were 10^{-4} cm, the collapse would take about $\frac{1}{10}$ sec, whereas for 10^{-3} cm size, the collapse of the state vector would take place in only some 10^{-6} sec. Also, this is when the lump is isolated from the environment; the decay is hastened by mass movement in the environment. Schemes to solve the measurement problem in QT of this sort tend to run into problems with energy conservation and locality. But in GR there is an inbuilt uncertainty to the energy of gravity, particularly with regard to how this would contribute to the superposed state. The energy of gravity is nonlocal in GR: gravitational potential energy contributes (negatively) nonlocally to the total energy, and gravitational waves can carry (positive) nonlocal energy away from a system. Even flat spacetime can contribute to the total energy in certain circumstances. The energy uncertainty in the superposed state of two mass locations, as considered here, is consistent (by Heisenberg's uncertainty) with the decay time (4.4).

QUESTIONS AND ANSWERS

Question: Professor Hawking mentioned that the gravitational field was in some way more special than the other fields. What do you think about this?

Answer: The gravitational field certainly is special. Somehow there is an irony in the history of the subject: Newton started physics with the theory of gravity and this theory was the original paradigm for all other physical interactions. But now it turns out that gravity is in fact distinctly different from other interactions. Gravity is the only one which affects causality, with profound implications with regard to black holes and information loss.

CHAPTER FIVE

Quantum Cosmology

S. W. Hawking

IN MY THIRD LECTURE, I shall turn to cosmology. Cosmology used to be considered a pseudoscience and the preserve of physicists who might have done useful work in their earlier years, but who had gone mystic in their dotage. There were two reasons for this. The first was that there was an almost total absence of reliable observations. Indeed, until the 1920s about the only important cosmological observation was that the sky at night is dark. But people didn't appreciate the significance of this. However, in recent years the range and quality of cosmological observations has improved enormously with developments in technology. So this objection against regarding cosmology as a science, that it doesn't have an observational basis, is no longer valid.

There is, however, a second and more serious objection. Cosmology cannot predict anything about the universe unless it makes some assumption about the initial conditions. Without such an assumption, all one can say is that things are as they are now because they were as they were at an earlier stage. Yet many people believe that science should be concerned only with the local laws which govern how the universe evolves in time. They would feel that the boundary conditions for the universe that determine how the universe began were a question for metaphysics or religion, rather than science.

The situation was made worse by the theorems that Roger and I proved. These showed that, according to general relativity, there should be a singularity in our past. At this singularity the field equations could not be defined. Thus classical general relativity brings about its own downfall: it predicts that it can't predict the universe.

Although many people welcomed this conclusion, it has always profoundly disturbed me. If the laws of physics could break down at the beginning of the universe, why couldn't they break down anywhere? In quantum theory it is a principle that anything can happen if it is not absolutely forbidden. Once one allows that singular histories could take part in the path integral, they could occur anywhere and predictability would disappear completely. If the laws of physics break down at singularities, they could break down anywhere.

The only way to have a scientific theory is if the laws of physics hold everywhere, including at the beginning of the universe. One can regard this as a triumph for the principles of democracy: Why should the beginning of the universe be exempt from the laws that apply to other points? If all points are equal, one can't allow some to be more equal than others.

To implement the idea that the laws of physics hold everywhere, one should take the path integral only over nonsingular metrics. One knows, in the ordinary path integral case, that the measure is concentrated on nondifferentiable paths. But these are the completion, in some suitable topology, of the set of smooth paths with well-defined action. Similarly, one would expect that the path integral for quantum gravity should be taken over the completion of the space of smooth metrics. What the path integral can't include is metrics with singularities whose action is not defined.

In the case of black holes, we saw that the path integral should be taken over Euclidean, that is, positive definite metrics. This meant that the singularities of black holes, like the Schwarzschild solution, did not appear on the Euclidean metrics, which did not go inside the horizon. Instead, the horizon was like the origin of polar coordinates. The action of the Euclidean metric was therefore well defined. One could regard this as a quantum version of cosmic censorship: the breakdown of the structure at a singularity should not affect any physical measurement.

It seems, therefore, that the path integral for quantum gravity should be taken over nonsingular Euclidean metrics. But what should the boundary conditions be on these metrics? There are two, and only two, natural choices. The first is metrics that approach the flat Eu-

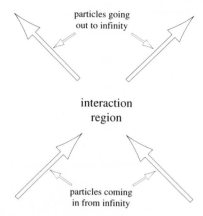

particles going
out to infinity

interaction
region

particles coming
in from infinity

Figure 5.1 In a scattering calculation, we make measurements on the incoming and outgoing particles at infinity, so we want to study asymptotically Euclidean metrics.

clidean metric outside a compact set. The second possibility is metrics on manifolds that are compact and without boundary.

Natural Choices for Path Integral for Quantum Gravity

1. Asymptotically Euclidean metrics.
2. Compact metrics without boundary.

The first class of asymptotically Euclidean metrics is obviously appropriate for scattering calculations (fig. 5.1). In these one sends particles in from infinity and observes what comes out again to infinity. All measurements are made at infinity, where one has a flat background metric and one can interpret small fluctuations in the fields as particles in the usual way. One doesn't ask what happens in the interaction region in the middle. That is why one does a path integral over all possible histories for the interaction region, that is, over all asymptotically Euclidean metrics.

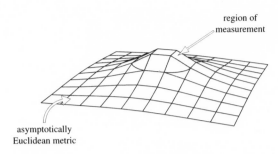

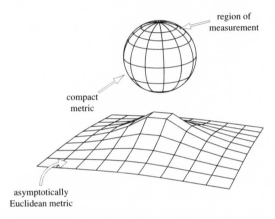

Figure 5.2 Cosmological measurements are made in a finite region, so we have to consider two types of asymptotically Euclidean metrics: connected ones (*top*) and disconnected ones (*bottom*).

However, in cosmology one is interested in measurements that are made in a finite region, rather than at infinity. We are on the inside of the universe, not looking in from the outside. To see what difference this makes, let us first suppose that the path integral for cosmology is to be taken over all asymptotically Euclidean metrics. Then there would be two contributions to probabilities for measurements in a finite region. The first would be from connected asymptotically Euclidean metrics. The second would be from disconnected metrics that consisted of a compact spacetime containing the region of measurements and a separate asymptotically Euclidean metric (fig. 5.2). One cannot exclude disconnected metrics from the path integral because they can be approximated by connected metrics in which the different components are joined by thin tubes or wormholes of negligible action.

Disconnected compact regions of spacetime won't affect scattering calculations because they aren't connected to infinity, where all measurements are made. But they will affect measurements in cosmology that are made in a finite region. Indeed, the contributions from such disconnected metrics will dominate over the contributions from connected asymptotically Euclidean metrics. Thus, even if one took the path integral for cosmology to be over all asymptotically Euclidean metrics, the effect would be almost the same as if the path integral had been over all compact metrics. It therefore seems more natural to take the path integral for cosmology to be over all compact metrics without boundary, as Jim Hartle and I proposed in 1983 (Hartle and Hawking 1983).

The No-Boundary Proposal (Hartle and Hawking)
The path integral for quantum gravity should be taken over all compact Euclidean metrics.

One can paraphrase this as "The Boundary Condition of the Universe Is That It Has No Boundary."

In the rest of this lecture I shall show that this no-boundary proposal seems to account for the universe we live in. That is, an isotropic and homogeneous expanding universe with small perturbations. We can observe the spectrum and statistics of these perturbations in the fluctuations in the microwave background. The results so far agree with the predictions of the no-boundary proposal. It will be a real test of the proposal and the whole Euclidean quantum gravity program when the observations of the microwave background are extended to smaller angular scales.

In order to use the no-boundary proposal to make predictions, it is useful to introduce a concept that can describe the state of the universe

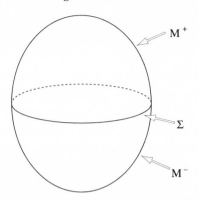

Figure 5.3 The surface Σ divides the compact, simply connected manifold M into two parts, M^+ and M^-.

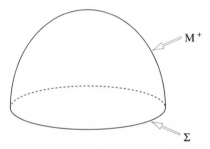

Figure 5.4 The wave function is given by a path integral over M^+.

at one time. Consider the probability that the spacetime manifold M contains an embedded three-dimensional manifold Σ with induced metric h_{ij}. This is given by a path integral over all metrics g_{ab} on M that induce h_{ij} on Σ.

$$\text{Probability of induced metric } h_{ij} \text{ on } \Sigma = \int_{\substack{\text{metrics on } M \text{ that} \\ \text{induce } h_{ij} \text{ on } \Sigma}} d[g]\, e^{-I}.$$

If M is simply connected, which I will assume, the surface Σ will divide M into two parts, M^+ and M^- (fig. 5.3). In this case, the probability for Σ to have the metric h_{ij} can be factorized. It is the product of two wave functions Ψ^+ and Ψ^-. These are given by path integrals over all metrics on M^+ and M^- respectively, that induce the given three-metric h_{ij} on Σ.

$$\text{Probability of } h_{ij} \quad = \quad \Psi^+(h_{ij}) \times \Psi^-(h_{ij}), \quad \text{where}$$

$$\Psi^+(h_{ij}) \quad = \quad \int_{\substack{\text{metrics on } M^+ \text{ that} \\ \text{induce } h_{ij} \text{ on } \Sigma}} d[g]\, e^{-I}.$$

In most cases, the two wave functions will be equal and I will drop the superscripts $+$ and $-$. Ψ is called the wave function of the universe. If there are matter fields ϕ, the wave function will also depend on their values ϕ_0 on Σ. But it will not depend explicitly on time because there is no preferred time coordinate in a closed universe. The no-boundary proposal implies that the wave function of the universe is given by a path integral over fields on a compact manifold M^+ whose only boundary is the surface Σ (fig. 5.4). The path integral is taken over all metrics and matter fields on M^+ that agree with the metric h_{ij} and matter fields ϕ_0 on Σ.

One can describe the position of the surface Σ by a function τ of three coordinates x_i on Σ. But the wave function defined by the path integral can't depend on τ or on the choice of the coordinates x_i. This implies that the wave function Ψ has to obey four functional differential equations. Three of these equations are called the *momentum constraints*.

Momentum Constraint Equations

$$\left(\frac{\partial \Psi}{\partial h_{ij}}\right)_{;j} = 0$$

They express the fact that the wave function should be the same for different three-metrics h_{ij} that can be obtained from each other by transformations of the coordinates x_i. The fourth equation is called the *Wheeler-DeWitt equation*.

Wheeler-DeWitt Equation

$$\left(G_{ijkl}\frac{\partial^2}{\partial h_{ij}\partial h_{kl}} - h^{\frac{1}{2}}\,^3R\right)\Psi = 0.$$

It corresponds to the independence of the wave function of τ. One can think of it as the Schrödinger equation for the universe. But there is no time-derivative term because the wave function does not depend on time explicitly.

In order to estimate the wave function of the universe, one can use the saddle point approximation to the path integral, as in the case of black holes. One finds a Euclidean metric g_0 on the manifold M^+ that satisfies the field equations and induces the metric h_{ij} on the boundary Σ. One can then expand the action in a power series around the background metric g_0.

$$I[g] = I[g_0] + \frac{1}{2}\delta g I_2 \delta g + \cdots$$

As before, the term linear in the perturbations vanishes. The quadratic term can be regarded as giving the contribution of gravitons on the background and the higher-order terms as interactions between the gravitons. These can be ignored when the radius of curvature of the background is large compared to the Planck scale. Therefore,

$$\Psi \approx \frac{1}{(\det I_2)^{\frac{1}{2}}}e^{-I[g_0]}.$$

One can see what the wave function is like from a simple example. Consider a situation in which there are no matter fields but there is

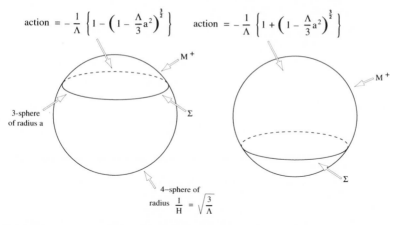

$$\text{action} = -\frac{1}{\Lambda}\left\{1 - \left(1 - \frac{\Lambda}{3}a^2\right)^{\frac{3}{2}}\right\} \qquad \text{action} = -\frac{1}{\Lambda}\left\{1 + \left(1 - \frac{\Lambda}{3}a^2\right)^{\frac{3}{2}}\right\}$$

Figure 5.5 The two possible Euclidean solutions M^+ with boundary Σ, and their actions.

a positive cosmological constant Λ. Let us take the surface Σ to be a three-sphere and the metric h_{ij} to be the round three-sphere metric of radius a. Then the manifold M^+ bounded by Σ can be taken to be the four-ball. The metric that satisfies the field equations is part of a four-sphere of radius $\frac{1}{H}$, where $H^2 = \frac{\Lambda}{3}$. The action is:

$$I = \frac{1}{16\pi}\int (R - 2\Lambda)(-g)^{\frac{1}{2}}d^4x + \frac{1}{8\pi}\int K(\pm h)^{\frac{1}{2}}d^3x.$$

For a three-sphere Σ of radius less than $\frac{1}{H}$ there are two possible Euclidean solutions: either M^+ can be less than a hemisphere or it can be more (fig. 5.5). However, there are arguments that show that one should pick the solution corresponding to less than a hemisphere.

The next figure (fig. 5.6) shows the contribution to the wave function that comes from the action of the metric g_0. When the radius of Σ is less than $\frac{1}{H}$, the wave function increases exponentially like e^{a^2}. However, when a is greater than $\frac{1}{H}$, one can analytically continue the result for smaller a and obtain a wave function that oscillates very rapidly.

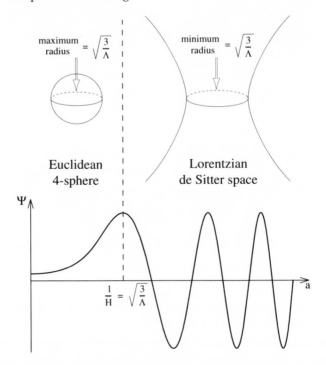

Figure 5.6 The wave function as a function of the radius of Σ.

One can interpret this wave function as follows. The real time solution of the Einstein equations with a Λ term and maximal symmetry is de Sitter space. This can be embedded as a hyperboloid in five-dimensional Minkowski space (see box 5.A). One can think of it as a closed universe that shrinks down from infinite size to a minimum radius and then expands again exponentially. The metric can be written in the form of a Friedmann universe with scale factor $\cosh Ht$. Putting $\tau = it$ converts the cosh into cos giving the Euclidean metric on a four-sphere of radius $\frac{1}{H}$ (see box 5.B). Thus, one gets the idea that a wave function which varies exponentially with the three-metric h_{ij} corresponds to an imaginary time Euclidean metric. On the other hand, a wave function which oscillates rapidly corresponds to a real-time Lorentzian metric.

Box 5.A. Lorentzian–de Sitter Metric

$$ds^2 = -dt^2 + \frac{1}{H^2}\cosh Ht(dr^2 + \sin^2 r(d\theta^2 + \sin^2\theta d\phi^2))$$

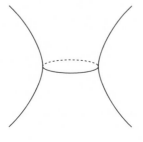

Box 5.B. Euclidean Metric

$$ds^2 = d\tau^2 + \frac{1}{H^2}\cos H\tau(dr^2 + \sin^2 r(d\theta^2 + \sin^2\theta d\phi^2))$$

As in the case of the pair creation of black holes, one can describe the spontaneous creation of an exponentially expanding universe. One joins the lower half of the Euclidean four-sphere to the upper half of the Lorentzian hyperboloid (fig. 5.7). Unlike the black hole pair creation, one couldn't say that the de Sitter universe was created out of field energy in a preexisting space. Instead, it would quite literally be created out of nothing: not just out of the vacuum, but out of absolutely nothing at all, because there is nothing outside the universe. In

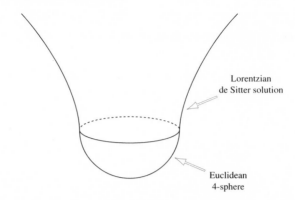

Figure 5.7 The tunneling to produce an expanding universe is described by joining half the Euclidean solution to half the Lorentzian solution.

the Euclidean regime, the de Sitter universe is just a closed space like the surface of the Earth but with two more dimensions. If the cosmological constant is small compared to the Planck value, the curvature of the Euclidean four-sphere should be small. This will mean that the saddle point approximation to the path integral should be good, and that the calculation of the wave function of the universe won't be affected by our ignorance of what happens at very high curvatures.

One can also solve the field equations for boundary metrics that aren't exactly the round three-sphere metric. If the radius of the three-sphere is less than $\frac{1}{H}$, the solution is a real Euclidean metric. The action will be real and the wave function will be exponentially damped compared to the round three-sphere of the same volume. If the radius of the three-sphere is greater than this critical radius, there will be two complex conjugate solutions and the wave function will oscillate rapidly with small changes in h_{ij}.

Any measurement made in cosmology can be formulated in terms of the wave function. Thus, the no-boundary proposal makes cosmology into a science, because one can predict the result of any observation. The case we have just been considering of no matter fields and only a cosmological constant does not correspond to the universe we live in. Nevertheless, it is a useful example, both because it is a simple model that can be solved fairly explicitly and because, as we shall see, it seems to correspond to the early stages of the universe.

Although it is not obvious from the wave function, a de Sitter universe has thermal properties rather like a black hole. One can see this by writing the de Sitter metric in a static form rather like the Schwarzschild solution (see box 5.C).

Box 5.C. Static Form of the de Sitter Metric

$$ds^2 = -(1 - H^2r^2)dt^2 + (1 - H^2r^2)^{-1}dr^2 + r^2(d\theta^2 + \sin^2\theta d\phi^2)$$

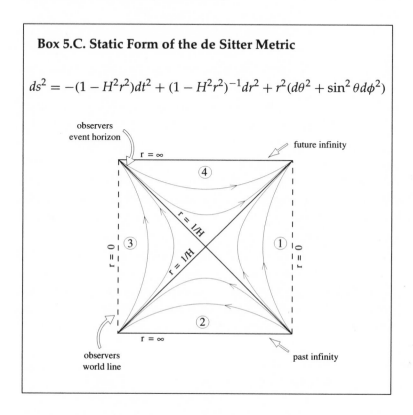

There is an apparent singularity at $r = \frac{1}{H}$. However, as in the Schwarzschild solution, one can remove it by a coordinate transformation and it corresponds to an event horizon. This can be seen from the Carter-Penrose diagram, which is a square. The dotted vertical line on the left represents the center of spherical symmetry where the radius r of the two-spheres goes to zero. Another center of spherical symmetry is represented by the dotted vertical line on the right. The horizontal lines at the top and bottom represent past and future

infinity, which are spacelike in this case. The diagonal line from top
left to bottom right is the boundary of the past of an observer at the
left-hand center of symmetry. Thus it can be called his event horizon.
However, an observer whose world line ends up at a different place
on future infinity will have a different event horizon. Thus event
horizons are a personal matter in de Sitter space.

If one returns to the static form of the de Sitter metric and puts
$\tau = it$, one gets a Euclidean metric. There is an apparent singularity
on the horizon. However, by defining a new radial coordinate and
identifying τ with period $\frac{2\pi}{H}$, one gets a regular Euclidean metric,
which is just the four-sphere. Because the imaginary time coordinate
is periodic, de Sitter space and all quantum fields in it will behave as
if they were at a temperature $\frac{H}{2\pi}$. As we shall see, we can observe the
consequences of this temperature in the fluctuations in the microwave
background. One can also apply arguments similar to the black hole
case to the action of the Euclidean–de Sitter solution. One finds that
it has an intrinsic entropy of $\frac{\pi}{H^2}$, which is a quarter of the area of the
event horizon. Again this entropy arises for a topological reason: the
Euler number of the four-sphere is two. This means that there cannot
be a global time coordinate on Euclidean–de Sitter space. One can
interpret this cosmological entropy as reflecting an observer's lack of
knowledge of the universe beyond his event horizon.

$$\text{Euclidean metric periodic with period } \frac{2\pi}{H}$$

$$\Rightarrow \left\{ \begin{array}{c} \text{Temperature} = \frac{H}{2\pi} \\ \text{Area of event horizon} = \frac{4\pi}{H^2} \\ \text{Entropy} = \frac{\pi}{H^2} \end{array} \right.$$

De Sitter space is not a good model of the universe in which we live,
because it is empty and it is expanding exponentially. We observe
that the universe contains matter, and we deduce from the microwave
background and the abundance of light elements that it must have

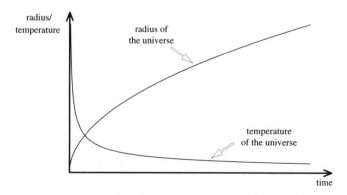

Figure 5.8 The radius and temperature of the universe as a function of time in the hot big bang model.

been much hotter and denser in the past. The simplest scheme that is consistent with our observations is called the "hot big bang" model (fig. 5.8). In this scenario, the universe starts at a singularity filled with radiation at an infinite temperature. As it expands, the radiation cools and its energy density goes down. Eventually, the energy density of the radiation becomes less than that of the density of the nonrelativistic matter, and the expansion becomes matter dominated. However, we can still observe the remains of the radiation in a background of microwave radiation at a temperature of about 3°K above absolute zero.

The trouble with the hot big bang model is the trouble with all cosmology that has no theory of initial conditions: it has no predictive power. Because general relativity would break down at a singularity, anything could come out of the big bang. So why is the universe so homogeneous and isotropic on a large scale, yet has local irregularities such as galaxies and stars? And why is the universe so close to the dividing line between collapsing again and expanding indefinitely? In order to be as close as we are now, the rate of expansion early on had to be chosen fantastically accurately. If the rate of expansion one second after the big bang had been less by one part in 10^{10}, the universe would have collapsed after a few million years. If it had been greater by one part in 10^{10}, the universe would have been essentially empty after a few million years. In neither case would it have lasted long enough for life to develop. Thus one either has to appeal to the anthropic principle or find some physical explanation of why the universe is the way it is.

Hot Big Bang model does not explain why:

1. The universe is nearly homogeneous and isotropic but with small perturbations.
2. The universe is expanding at almost exactly the critical rate to avoid collapsing again.

Some people have claimed that what is called *inflation* removes the need for a theory of initial conditions. The idea is that the universe could start out at the big bang in almost any state. In those parts of the universe in which conditions were suitable there would be a period of exponential expansion called inflation. Not only could this increase the size of the region by an enormous factor of 10^{30} or more, it would also leave the region homogeneous and isotropic and expanding at just the critical rate to avoid collapsing again. The claim would be that intelligent life would develop only in regions that inflated. We should not, therefore, be surprised that our region is homogeneous and isotropic and is expanding at just the critical rate.

However, inflation alone cannot explain the present state of the universe. One can see this by taking any state for the universe now and running it back in time. Providing it contains enough matter, the singularity theorems will imply that there was a singularity in the past. One can choose the initial conditions of the universe at the big bang to be the initial conditions of this model. In this way, one can show that arbitrary initial conditions at the big bang can lead to any state now. One can't even argue that most initial states lead to a state like we observe today: the natural measure of both the initial conditions that do lead to a universe like ours and those that don't is infinite. One can't therefore claim that one is bigger than the other.

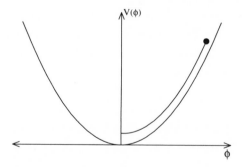

Figure 5.9 The potential for a massive scalar field.

On the other hand, we saw, in the case of gravity with a cosmological constant but no matter fields, that the no-boundary condition could lead to a universe that was predictable within the limits of quantum theory. This particular model did not describe the universe in which we live, which is full of matter and has zero or very small cosmological constant. However, one can get a more realistic model by dropping the cosmological constant and including matter fields. In particular, one seems to need a scalar field ϕ with potential $V(\phi)$. I shall assume that V has a minimum value of zero at $\phi = 0$. A simple example would be a massive scalar field $V = \frac{1}{2}m^2\phi^2$ (fig. 5.9).

Energy-Momentum Tensor of a Scalar Field

$$T_{ab} = \phi_{,a}\phi_{,b} - \frac{1}{2}g_{ab}\phi_{,c}\phi^{,c} - g_{ab}V(\phi).$$

One can see from the energy-momentum tensor that if the gradient of ϕ is small, $V(\phi)$ acts like an effective cosmological constant.

The wave function will now depend on the value ϕ_0 of ϕ on Σ, as well as on the induced metric h_{ij}. One can solve the field equations for small round three-sphere metrics and large values of ϕ_0. The solution with that boundary is approximately part of a four-sphere and a nearly constant ϕ field. This is like the de Sitter case with the potential $V(\phi_0)$ playing the role of the cosmological constant. Similarly, if the radius a of the three-sphere is a bit bigger than the radius of the Euclidean four-sphere, there will be two complex conjugate solutions. These will be like half of the Euclidean four-sphere joined onto a Lorentzian–de Sitter solution with almost constant ϕ. Thus the no-boundary proposal predicts the spontaneous creation of an exponentially expanding universe in this model as well as in the de Sitter case.

One can now consider the evolution of this model. Unlike the de Sitter case, it will not continue indefinitely with exponential expansion. The scalar field will run down the hill of the potential V to the minimum at $\phi = 0$. However, if the initial value of ϕ is larger than the Planck value, the rate of rolldown will be slow compared to the expansion timescale. Thus the universe will expand almost exponentially by a large factor. When the scalar field gets down to order one, it will start to oscillate about $\phi = 0$. For most potentials V, the oscillations will be rapid compared to the expansion time. It is normally assumed that the energy in these scalar field oscillations will be converted into pairs of other particles and will heat up the universe. This, however, depends on an assumption about the arrow of time. I shall come back to this shortly.

The exponential expansion by a large factor would have left the universe with almost exactly the critical rate of expansion. Thus the no-boundary proposal can explain why the universe is still so close to the critical rate of expansion. To see what it predicts for the homogeneity and isotropy of the universe, one has to consider three-metrics h_{ij} which are perturbations of the round three-sphere metric. One can expand these in terms of spherical harmonics. There are three kinds: scalar harmonics, vector harmonics, and tensor harmonics. The vector harmonics just correspond to changes of the coordinates x_i on successive three-spheres and play no dynamical role. The tensor harmonics correspond to gravitational waves in the expanding universe, while the scalar harmonics correspond partly to coordinate freedom and partly to density perturbations.

Tensor harmonics—Gravitational waves

Vector harmonics—Gauge

Scalar harmonics—Density perturbations

One can write the wave function Ψ as a product of a wave function Ψ_0 for a round three-sphere metric of radius a times wave functions for the coefficients of the harmonics:

$$\Psi[h_{ij}, \phi_0] = \Psi_0(a, \bar{\phi})\Psi_a(a_n)\Psi_b(b_n)\Psi_c(c_n)\Psi_d(d_n)$$

One can then expand the Wheeler-DeWitt equation for the wave function to all orders in the radius a and the average scalar field $\bar{\phi}$, but to first order in the perturbations. One gets a series of Schrödinger equations for the rate of change of the perturbation wave functions with respect to the time coordinate of the background metric.

Schrödinger Equations

$$i\frac{\partial\Psi(d_n)}{\partial t} = \frac{1}{2a^3}\left(-\frac{\partial^2}{\partial d_n^2} + n^2 d_n^2 a^4\right)\Psi(d_n), \quad \text{etc.}$$

One can use the no-boundary condition to obtain initial conditions for the perturbation wave functions. One solves the field equations for a small but slightly distorted three-sphere. This gives the perturbation wave function in the exponentially expanding period. One can then evolve it using the Schrödinger equation.

The tensor harmonics that correspond to gravitational waves are the simplest to consider. They don't have any gauge degrees of freedom and they don't interact directly with the matter perturbations. One can use the no-boundary condition to solve for the initial wave function of the coefficients d_n of the tensor harmonics in the perturbed metric.

Ground State

$$\Psi(d_n) \propto e^{-\frac{1}{2}na^2d_n^2} = e^{-\frac{1}{2}\omega x^2},$$

where $x = a^{\frac{3}{2}}d_n$ and $\omega = \dfrac{n}{a}$

One finds that it is the ground-state wave function for a harmonic oscillator at the frequency of the gravitational waves. As the universe expands, the frequency will fall. While the frequency is greater than the expansion rate $\dot{a}/a$, the Schrödinger equation will allow the wave function to relax adiabatically and the mode will remain in its ground state. Eventually, however, the frequency will become less than the expansion rate, which is roughly constant during the exponential expansion. When this happens, the Schrödinger equation will no longer be able to change the wave function fast enough that it can remain in the ground state while the frequency changes. Instead, it will freeze in the shape it had when the frequency fell below the expansion rate.

After the end of the exponential expansion era, the expansion rate will decrease faster than the frequency of the mode. This is equivalent to saying that an observer's event horizon, the reciprocal of the expansion rate, increases faster than the wavelength of the mode. Thus the wavelength will get longer than the horizon during the inflation period and will come back within the horizon later on (fig. 5.10).

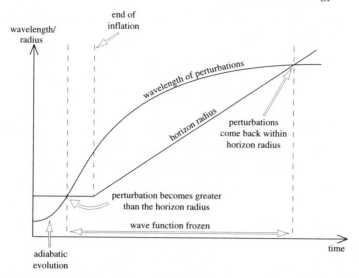

Figure 5.10 The wavelength and horizon radius as a function of time in inflation.

When it does, the wave function will still be the same as when the wave function froze. The frequency, however, will be much lower. The wave function will therefore correspond to a highly excited state, rather than to the ground state as it did when the wave function froze. These quantum excitations of the gravitational wave modes will produce angular fluctuations in the microwave background whose amplitude is the expansion rate (in Planck units) at the time the wave function froze. Thus, the COBE observations of fluctuations of one part in 10^5 in the microwave background place an upper limit of about 10^{-10} in Planck units on the energy density when the wave function froze. This is sufficiently low that the approximations I have used should be accurate.

However, the gravitational wave tensor harmonics give only an upper limit on the density at the time of freezing. The reason is that it turns out that the scalar harmonics give a larger fluctuation in the microwave background. There are two scalar harmonic degrees of freedom in the three-metric h_{ij}, and one in the scalar field. However, two of these scalar degrees correspond to coordinate freedom. Thus, there is only one physical scalar degree of freedom, and it corresponds to density perturbations.

The analysis for the scalar perturbations is very similar to that for the tensor harmonics if one uses one coordinate choice for the period up to the wave function freezing and another after that. In converting from one coordinate system to the other, the amplitudes get multiplied by a factor of the expansion rate divided by the average rate of change of ϕ. This factor will depend on the slope of the potential, but will be at least ten for reasonable potentials. This means the fluctuations in the microwave background that the density perturbations produce will be at least ten times bigger than those from the gravitational waves. Thus the upper limit on the energy density at the time of wave function freezing is only 10^{-12} of the Planck density. This is well within the range of the validity of the approximations I have been using. Thus, it seems we don't need string theory even for the beginning of the universe.

The spectrum of the fluctuations with angular scale agrees, within the accuracy of the present observations, with the prediction that it should be almost scale free. And the size of the density perturbations is just that required to explain the formation of galaxies and stars. Thus, it seems the no-boundary proposal can explain all the structure of the universe, including little inhomogeneities like ourselves.

COBE predictions plus gravitational wave perturbations	$\Rightarrow$	upper limit on energy density 10^{-10} Planck density
plus density perturbations	$\Rightarrow$	upper limit on energy density 10^{-12} Planck density
intrinsic gravitational temperature of early universe	$\approx$	10^{-6} Planck temperature $= 10^{26}$ degrees

One can think of the perturbations in the microwave background as arising from thermal fluctuations in the scalar field ϕ. The inflationary

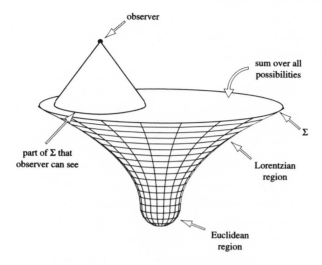

Figure 5.11 An observer can see only part of any surface Σ.

period has a temperature of the expansion rate over 2π because it is approximately periodic in imaginary time. Thus, in a sense, we don't need to find a little primordial black hole: we have already observed an intrinsic gravitational temperature of about 10^{26} degrees, or 10^{-6} of the Planck temperature.

What about the intrinsic entropy associated with the cosmological event horizon? Can we observe this? I think we can, and I think that it corresponds to the fact that objects like galaxies and stars are classical objects, even though they are formed by quantum fluctuations. If one looks at the universe on a spacelike surface Σ that spans the whole universe at one time, then it is in a single quantum state described by the wave function Ψ. However, we can never see more than half of Σ, and we are completely ignorant of what the universe is like beyond our past light cone. This means that in calculating the probability for observations, we have to sum over all possibilities for the part of Σ we don't observe (fig. 5.11). The effect of the summation is to change the part of the universe we observe from a single quantum state to what is called a *mixed state*, a statistical ensemble of different possibilities. Such decoherence, as it is called, is necessary if a system is to behave in a classical manner rather than a quantum one. People normally try to account for decoherence by interactions with an external system, such as a heat bath, that is not measured. In the case of the universe there is no external system, but I would suggest that the reason we observe classical behavior is that we can see only part of the universe. One might think that at late times one would be able to see all the universe and the event horizon would disappear. But this is not the case. The no-boundary proposal implies that the universe is spatially closed. A closed universe will collapse again before an observer has time to see all the universe. I have tried to show that the entropy of such a universe would be a quarter of the area of the event horizon at the time of maximum expansion (fig. 5.12). However, at the moment, I seem to be getting a factor of $\frac{3}{16}$ rather than a $\frac{1}{4}$. Obviously I'm either on the wrong track or I'm missing something.

I will end this lecture on a topic on which Roger and I have very different views—the arrow of time. There is a very clear distinction between the forward and backward directions of time in our region of the universe. One only has to watch a film being run backward to see the difference. Instead of cups falling off tables and getting broken, they would mend themselves and jump back on the table. If only real life were like that.

The local laws that physical fields obey are time symmetric, or more precisely, CPT invariant. Thus, the observed difference between the past and the future must come from the boundary conditions of the universe. Let us take it that the universe is spatially closed and that it expands to a maximum size and collapses again. As Roger has emphasized, the universe will be very different at the two ends of this history. At what we call the beginning of the universe, it seems to have been very smooth and regular. However, when it collapses again, we expect it to be very disordered and irregular. Because there are so many more disordered configurations than ordered ones, this means that the initial conditions would have had to be chosen incredibly precisely.

It seems, therefore, that there must be different boundary conditions at the two ends of time. Roger's proposal is that the Weyl tensor

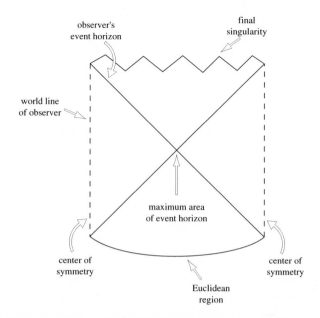

Figure 5.12 The universe will collapse to the final singularity before the observer can see the whole of the universe.

should vanish at one end of time but not the other. The Weyl tensor is that part of the curvature of spacetime that is not locally determined by the matter through the Einstein equations. It would have been small in the smooth, ordered early stages, but large in the collapsing universe. Thus this proposal would distinguish the two ends of time and so might explain the arrow of time (fig. 5.13).

I think Roger's proposal is Weyl in more than one sense of the word. First, it is not CPT invariant. Roger sees this as a virtue, but I feel one should hang on to symmetries unless there are compelling reasons to give them up. As I shall argue, it is not necessary to give up CPT. Second, if the Weyl tensor had been exactly zero in the early universe, it would have been exactly homogeneous and isotropic and would have remained so for all time. Roger's Weyl hypothesis could not explain the fluctuations in the background nor the perturbations that gave rise to galaxies and bodies like ourselves.

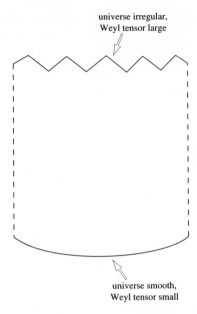

Figure 5.13 The Weyl tensor hypothesis for distinguishing the two ends of the universe.

Objections to Weyl Tensor Hypothesis

1. Not CPT invariant.
2. Weyl tensor cannot have been exactly zero. Doesn't explain small fluctuations.

Despite all this, I think Roger has put his finger on an important difference between the two ends of time. But the fact that the Weyl tensor was small at one end should not be imposed as an ad hoc boundary condition, but should be deduced from a more fundamental principle, the no-boundary proposal. As we have seen, this implies that perturbations about half the Euclidean four-sphere joined to half the Lorentzian–de Sitter solution are in their ground state. That is, they are as small as they can be, consistent with the uncertainty principle. This then would imply Roger's Weyl tensor condition: the Weyl tensor wouldn't be exactly zero but it would be as near to zero as it could be.

At first I thought that these arguments about perturbations being in their ground state would apply at both ends of the expansion-contraction cycle. The universe would start smooth and ordered and would get more disordered and irregular as it expanded. However, I thought it would have to return to a smooth and ordered state as it got smaller. This would have implied that the thermodynamic arrow of time would have to reverse in the contracting phase. Cups would mend themselves and jump back on the table. People would get younger, not older, as the universe got smaller again. It is not much good waiting for the universe to collapse again to return to our youth because it will take too long. But if the arrow of time reverses when the universe contracts, it might also reverse inside black holes. However, I wouldn't recommend jumping into a black hole as a way of prolonging one's life.

I wrote a paper claiming that the arrow of time would reverse when the universe contracted again. But after that, discussions with Don Page and Raymond Laflamme convinced me that I had made my greatest mistake, or at least my greatest mistake in physics: the universe would not return to a smooth state in the collapse. This would mean that the arrow of time would not reverse. It would continue pointing in the same direction as in the expansion.

How can the two ends of time be different? Why should perturbations be small at one end but not the other? The reason is there are two possible complex solutions of the field equations that match on to a small three-sphere boundary. One is as I have described earlier: it is approximately half the Euclidean four-sphere joined to a small part of the Lorentzian–de Sitter solution (fig. 5.14). The other possible solution has the same half-Euclidean four-sphere joined to a Lorentzian solution that expands to a very large radius and then contracts again to the small radius of the given boundary (fig. 5.15). Obviously, one solution corresponds to one end of time and the other to the other. The difference between the two ends comes from the fact that perturbations in the three-metric h_{ij} are heavily damped in the case of

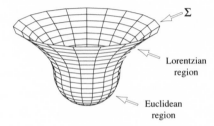

Figure 5.14 Half a Euclidean four-sphere joined on to a small Lorentzian region.

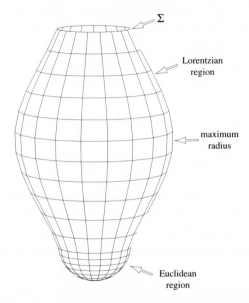

Figure 5.15 Half a Euclidean four-sphere joined on to a Lorentzian region that expands to maximum radius and then shrinks again.

the first solution with only a short Lorentzian period. However, the perturbations can be very large without being significantly damped in the case of the solution that expands and contracts again. This gives rise to the difference between the two ends of time that Roger has pointed out. At one end, the universe was very smooth and the Weyl tensor was very small. It could not, however, be exactly zero, for that would have been a violation of the uncertainty principle. Instead there would have been small fluctuations that later grew into galaxies and bodies like us. By contrast, the universe would have been very irregular and chaotic at the other end of time with a Weyl tensor that was typically large. This would explain the observed arrow of time and why cups fall off tables and break rather than mend themselves and jump back on.

As the arrow of time is not going to reverse—and as I have gone over time—I better draw my lecture to a close. I have emphasized what I consider the two most remarkable features that I have learned in my research on space and time: (1) that gravity curls up spacetime so that it has a beginning and an end; (2) that there is a deep connection between gravity and thermodynamics that arises because gravity itself determines the topology of the manifold on which it acts.

The positive curvature of spacetime produced singularities at which classical general relativity broke down. Cosmic censorship may shield us from black hole singularities but we see the big bang in full frontal nakedness. Classical general relativity cannot predict how the universe will begin. However, quantum general relativity, together with the no-boundary proposal, predicts a universe like the one we observe and even seems to predict the observed spectrum of fluctuations in the microwave background. However, although the quantum theory restores the predictability that the classical theory lost, it does not do so completely. Because we cannot see the whole of spacetime on account of black hole and cosmological event horizons, our observations are described by an ensemble of quantum states rather than by a single state. This introduces an extra level of unpredictability, but it may also be why the universe appears classical. This would rescue Schrödinger's cat from being half alive and half dead.

To have removed predictability from physics and then to have put it back again, but in a reduced sense, is quite a success story. I rest my case.

CHAPTER SIX

The Twistor View of Spacetime

R. Penrose

LET ME START with some remarks about Stephen's last lecture.

- **Classicality of Cats.** Stephen argued that because a certain region of spacetime is inaccessible we are forced into the density matrix description. However, this is not sufficient to explain the classical nature of observations in our region. The density matrix that corresponds to finding either a live cat |live⟩ or a dead cat |dead⟩ is the same density matrix that describes the mixture of the two superpositions,

$$\frac{1}{\sqrt{2}}(|\text{live}\rangle + |\text{dead}\rangle)$$

and

$$\frac{1}{\sqrt{2}}(|\text{live}\rangle - |\text{dead}\rangle).$$

Thus the density matrix alone does not say whether we see either a live or a dead cat or else one of these two superpositions. As I tried to argue at the end of my last lecture, we need more.

- **Weyl Curvature Hypothesis (WCH).** From what I understand of Stephen's position, I don't think that our disagreement is very great on this point. For an initial singularity the Weyl curvature is approximately zero and final ones have large Weyl curvature. Stephen argued that there must be small quantum fluctuations in the initial state and thus pointed out that the hypothesis that the initial Weyl curvature would be exactly zero could not be reasonable. I don't think that this is really a disagreement. The statement that the Weyl curvature is zero at the initial singularity is classical, and there is certainly some flexibility as to the precise statement of the hypothesis. Small perturbations are acceptable from my point of view, certainly in the quantum regime. We just need something to constrain it very near to zero. One would also expect thermal fluctuations in the Ricci tensor (due to matter) in the early universe, and maybe these would ultimately lead to the formation of $10^6 M_s$ black holes through Jeans instability. The vicinity of the singularities in these black holes would then have large Weyl curvature, but these are final-type rather than initial-type singularities, which is consistent with WCH.

I agree with Stephen that WCH is "botanic," i.e., phenomenological rather than explanatory. It needs an underlying theory to explain it. Maybe the "no-boundary proposal" (NBP) of Hartle and Hawking is a good candidate for the structure of the *initial* state. However, it seems to me that we need something very different to cope with the *final* state. In particular, a theory that explains the structure of singularities would have to violate T, PT, CT, and CPT in order that something of the nature of WCH can arise. This failure of time-symmetry might be quite subtle; it would have to be implicit in the rules of that theory which goes beyond QM. Stephen has argued that in view of a well-known theorem of QFT one should expect the theory to be CPT invariant. However, the proof of this theorem assumes that the usual rules of QFT apply and that the background space is flat. I think that both Stephen and I agree that the second condition doesn't hold and I also believe that the first assumption fails.

It also seems to me that the viewpoint that Stephen is proposing with regard to NBP does not imply that there are no white holes. If I understand Stephen's point of view correctly, then the NBP implies that there are essentially two solutions: one (A) where the perturbations increase away from the singularity, and one (B) where they die out. (A) corresponds essentially to the big bang, whereas (B) describes black hole singularities and the big crunch. The arrow of time, determined by the second law of thermodynamics, goes from an (A) solution to a (B) solution. However, I don't see how this interpretation of NBP excludes (B)-type white holes. On a separate issue, I worry about the "Euclideanization procedure."

Stephen's argument relies on the fact that one can glue a Euclidean and a Lorentzian solution together. However, there are only very few spaces for which one can do this, since it is required that they have both a Euclidean and a Lorentzian section. The generic case is certainly very far from that.

TWISTORS AND TWISTOR SPACE

What is it that really underlies the utility of Euclideanization in QFT? QFT requires a splitting of the field quantities into positive and negative frequency parts. The former propagate forward in time, the latter backward. To obtain the propagators of the theory, one needs a way of picking out the positive frequency (i.e., positive energy) part. A (different) framework for accomplishing this splitting is *twistor theory*—in fact, this splitting was one of the important original motivation for twistors (see Penrose 1986).

To explain this in detail, let us first consider complex numbers, fundamental to quantum theory, and whose structure, we shall find, also underlies spacetime structure. These are the numbers of the form $z = x + iy$, with x, y real, where i satisfies $i^2 = -1$, and the set of such numbers is denoted by $\mathbb{C}$. One can represent these numbers on a plane (the complex plane), or if a point at infinity is added, on a sphere—the *Riemann sphere*. This sphere is a very useful concept in many areas of mathematics such as analysis and geometry, but also in physics. The sphere can be projected onto a plane (together with a point at infinity). Take the plane through the equator of the sphere and join any point on the sphere to the South Pole. The point where this line intersects the plane is the corresponding point on the plane. Note that under this map the North Pole goes to the origin, the South Pole to infinity, and the real axis is mapped to a vertical circle going through the North and the South Poles. We can rotate the sphere so that the real numbers correspond to the equator, and I want to adopt this convention for the moment (see fig. 6.1).

Suppose we are given a complex-valued function $f(x)$ of a real variable x. By the above, we can think of f as being a function de-

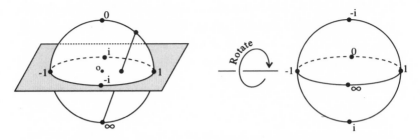

Figure 6.1 The Riemann sphere, representing all the complex numbers, together with ∞.

fined on the equator. The advantage of this point of view is that there is a natural criterion to decide whether f is positive or negative frequency: $f(x)$ is a positive frequency function if it can be extended to a holomorphic (complex analytic) function on the Northern Hemisphere, and similarly f is a negative frequency function if it can be extended likewise to the Southern Hemisphere. A general function can be split into a positive and negative frequency part. The idea of twistor theory is to use this device on spacetime itself in a global way. Given a field on Minkowski spacetime we want to split it, similarly, into positive and negative frequency parts. As a route to understanding this splitting, we shall construct twistor space. (See Penrose and Rindler 1986 and Huggett and Tod 1985 for more information about twistors).

Before doing this in detail, let us consider two important roles of the Riemann sphere in physics.

1. The wave function of a spin-$\frac{1}{2}$ particle may be in a linear superposition of "up" and "down":

$$w|\uparrow\rangle + z|\downarrow\rangle.$$

This state can be represented by a point z/w on the Riemann sphere, and this point corresponds to where the positive axis of the spin, taken out from the center, intersects the sphere. (For higher spin there is a more complicated construction, due

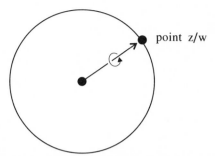

Figure 6.2 The space of spin directions for a spin-1/2 particle is the Riemann sphere of the ratio z/w of the amplitudes w (spin up) and z (spin down).

originally to Majorana 1932; cf. also Penrose 1994, which still uses the Riemann sphere.) This relates the complex amplitudes of QM to spacetime structure (fig. 6.2).

2. Imagine an observer situated at a point in spacetime, out in space looking at the stars. Suppose she plots the angular position of these stars on a sphere. Now, if a second observer were to pass through the same point at the same time, but with a velocity relative to the first observer then, owing to abberation effects, he would map the stars in different positions on the sphere. What is remarkable is that the different positions of the points on the sphere are related by a special transformation called a *Möbius transformation*. Such transformations form precisely the group that preserves the complex structure of the Riemann sphere. Thus, the space of light rays through a spacetime point is, in a natural sense, a Riemann sphere. I find it very beautiful, moreover, that the fundamental symmetry group of physics relating observers with different velocities, the (restricted) Lorentz group, can be realized as the automorphism group of the simplest one-(complex-)dimensional manifold, the Riemann sphere (see fig. 6.3 and Penrose and Rindler 1984).

The basic idea of twistor theory is to try to exploit this link between QM and spacetime structure—as manifested in the Riemann

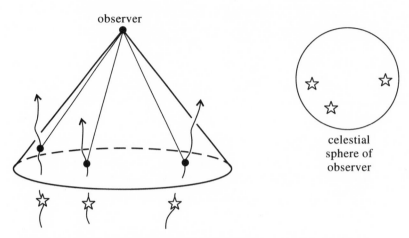

observer

celestial
sphere of
observer

Figure 6.3 The celestial sphere of an observer, in relativity theory, is naturally a Riemann sphere.

sphere—by extending this idea to the whole of spacetime. We shall try to regard entire light rays as more fundamental even than spacetime points. In this way, we consider spacetime to be a secondary concept and regard twistor space—initially the space of light rays—as the more fundamental space. These two spaces are related by a correspondence that represents light rays in spacetime as points in twistor space. A point in spacetime is then represented by the set of light rays that passes through it. Thus a point in spacetime becomes a Riemann sphere in twistor space. We should think of twistor space as the space in terms of which we should describe physics (fig. 6.4).

As I've presented twistor space so far it has five (real) dimensions and thus will not be a complex space, as complex spaces are always even (real) dimensional. If we think of light rays as photon histories, we also need to take into account the energy of the photon and also its helicity, which can be left- or right-handed. This is a little more complicated than just a light ray, but the virtue of this is that we end up with a complex projective three-space (six real dimensions), $\mathbb{CP}_3$. This is *projective twistor space* ($\mathbb{PT}$). It has a five-dimensional subspace $\mathbb{PN}$ which splits the space $\mathbb{PT}$ into two parts, the left- and right-handed pieces $\mathbb{PT}^-$ and $\mathbb{PT}^+$.

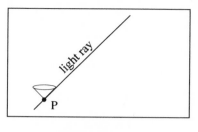
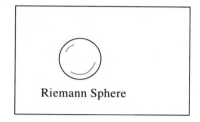

space-time   (projective) twistor space

Figure 6.4 In the basic twistor correspondence, light rays in (Minkowski) spacetime are represented as points in (projective) twistor space, and spacetime points are represented as Riemann spheres.

Now, points in spacetime are given by four real numbers, and projective twistor space can be coordinatized by the ratios of four complex numbers. If a light ray, represented by (Z^0, Z^1, Z^2, Z^3) in twistor space, goes through the point (r^0, r^1, r^2, r^3) in spacetime, then the *incidence* relation

$$\begin{pmatrix} Z^0 \\ Z^1 \end{pmatrix} = \frac{i}{\sqrt{2}} \begin{pmatrix} r^0 + r^3 & r^1 + ir^2 \\ r^1 - ir^2 & r^0 - r^3 \end{pmatrix} \begin{pmatrix} Z^2 \\ Z^3 \end{pmatrix} \tag{6.1}$$

is satisfied. The incidence relation (6.1) provides the basis of the twistor correspondence.

I shall need to introduce some two-spinor notation. This is usually where people start to get confused, but for calculations of any detail, this notation is extremely handy. For any four-vector r^a define the quantity $r^{AA'}$, whose matrix of components is given by

$$r^{AA'} = \begin{pmatrix} r^{00'} & r^{01'} \\ r^{10'} & r^{11'} \end{pmatrix} = \frac{1}{\sqrt{2}} \begin{pmatrix} r^0 + r^3 & r^1 + ir^2 \\ r^1 - ir^2 & r^0 - r^3 \end{pmatrix}.$$

The condition that r^a be *real* is simply that $r^{AA'}$ be *Hermitian*. A point in twistor space is defined by two spinors, with components

$$\omega^A \equiv \begin{pmatrix} \omega^1 \\ \omega^2 \end{pmatrix} = \begin{pmatrix} Z^0 \\ Z^1 \end{pmatrix} \quad \pi_{A'} \equiv \begin{pmatrix} \pi'_0 \\ \pi'_1 \end{pmatrix} = \begin{pmatrix} Z^2 \\ Z^3 \end{pmatrix}.$$

The incidence relation (6.1) then becomes

$$\omega = ir\pi.$$

It should be noted that under a shift of origin, according to which r^a is replaced

$$r^a \mapsto r^a - Q^a,$$

we have

$$\omega^A \mapsto \omega^A - iQ^{AA'}\pi_{A'},$$

whereas $\pi_{A'}$ remains unchanged:

$$\pi_{A'} \mapsto \pi_{A'}.$$

The twistor represents the four components of momentum p_a (three of which are independent) and the six components of angular momentum M^{ab} (four being independent of these) of a massless particle. The expressions are

$$p_{AA'} = i\bar{\pi}_A \pi_{A'}, \quad M^{AA'BB'} = i\omega^{(A}\bar{\pi}^{B)}\epsilon^{A'B'} - i\epsilon^{AB}\bar{\omega}^{(A'}\pi^{B')},$$

where parentheses denote the symmetric part and ϵ^{AB} and $\epsilon^{A'B'}$ are the skew Levi-Civita symbols. These expressions incorporate the fact that the momentum p_a is null and future pointing, and that the Pauli-Lubanski spin vector is the helicity s times the four-momentum. These quantities determine the twistor variables $(\omega^A, \pi_{A'})$ up to an overall twistor phase multiplier. The helicity can be written

$$s = \frac{1}{2}Z^\alpha \bar{Z}_\alpha,$$

where the complex conjugate of the twistor $Z^\alpha = (\omega^A, \pi_{A'})$ is the *dual* twistor $\bar{Z}_\alpha = (\bar{\pi}_A, \bar{\omega}^{A'})$. (Note that complex conjugation interchanges primed and unprimed spinor indices, and it interchanges twistors with their duals.) Here, $s > 0$ corresponds to right-handed particles and thus to what we refer to as the upper half of twistor space $\mathbb{PT}^+$ and $s < 0$ to left-handed particles, i.e., to the lower half $\mathbb{PT}^-$. It is in the case $s = 0$ that we get actual light rays. (The equation for $\mathbb{PN}$, the space of light rays, is therefore $Z^\alpha \bar{Z}_\alpha = 0$, i.e., $\omega^A \bar{\pi}_A + \pi_{A'} \bar{\omega}^{A'} = 0$.)

QUANTIZED TWISTORS

We wish to have a quantum theory of twistors, and for this we need to define a twistor wave function, a complex-valued function $f(Z^\alpha)$, on twistor space. *Any* function $f(Z^\alpha)$ is not a priori a wave function, as Z^α includes components involving position variables as well as all the momentum variables, and we cannot use all of these at the same time in a wave function. Position and momentum do not commute. In twistor space the commutation relations are

$$[Z^\alpha, \bar{Z}_\beta] = \hbar \delta_\beta^\alpha \quad [Z^\alpha, Z^\beta] = 0 \quad [\bar{Z}_\alpha, \bar{Z}_\beta] = 0.$$

Thus Z^α and $\bar{Z}_\alpha$ are conjugate variables, and the wave function must be a function of one only and not the other. This means the wave function must be a holomorphic (or else an antiholomorphic) function of Z^α.

We must now check how the previous expressions depend on the operator ordering. It turns out that the expressions for momentum and angular momentum are independent of the ordering and thus canonically determined. On the other hand, the expression for the helicity depends on the ordering, and we have to take the correct definition. For this we must take the symmetric product, i.e.,

$$s = \frac{1}{4} \left(Z^\alpha \bar{Z}_\alpha + \bar{Z}_\alpha Z^\alpha \right),$$

which, in the Z^α-space picture, can be reexpressed as

$$
s = \frac{\hbar}{2}\left(-2 - Z^\alpha \frac{\partial}{\partial Z^\alpha}\right)
$$

$$
= \frac{\hbar}{2}(-2 - \text{degree of homogeneity in } Z^\alpha).
$$

We can decompose a wave function into eigenstates of s. These are then precisely the wave functions of definite homogeneity. For example, a spinless particle with zero helicity has a twistor wave function of homogeneity -2. A left-handed spin-$\frac{1}{2}$ particle has helicity $s = -\frac{\hbar}{2}$ and therefore has a twistor wave function with homogeneity -1, whereas a right-handed version of such a particle (helicity $s = \frac{\hbar}{2}$) would have a twistor wave function of homogeneity -3. For spin 2, the right- and left-handed twistor wave functions have respective homogeneities -6 and $+2$.

This may look a little lopsided, as after all GR is left-right symmetric. But this may not be such a bad thing, as Nature herself is left-right asymmetric. Furthermore, the Ashtekar "new variables," which are very powerful tools in GR, are also left-right asymmetric. It is interesting that we are led to this left-right asymmetry in these different ways.

One might think that we can restore the symmetry by changing $Z^\alpha \leftrightarrow \bar{Z}_\alpha$, reversing the table of homogeneities and then using Z^α for one helicity and $\bar{Z}_\alpha$ for the other. However, just as we cannot mix position- and momentum-space pictures simultaneously in ordinary QT in this way, similarly we cannot mix Z^α and $\bar{Z}_\alpha$ pictures. We must choose one or the other. Whether one or the other is more fundamental remains to be seen.

Next we want to obtain a spacetime description of $f(Z)$. This is done via a contour integral

$$
\left\{ \begin{array}{c} \phi_{A'\cdots G'}(r) \\ \text{or} \\ \phi_{A\cdots G}(r) \end{array} \right\} = \int_{\omega = ir\pi} \left\{ \begin{array}{c} \pi_{A'}\cdots\pi_{G'} \\ \text{or} \\ \dfrac{\partial}{\partial\omega^A}\cdots\dfrac{\partial}{\partial\omega^G} \end{array} \right\} f(Z^\alpha)\pi_{E'}\,d\pi^{E'},
$$

where the integral is over a contour in the space of those Zs incident with r (recall Z has two parts ω and π) and the number of πs or $\partial/\partial\omega$s depends on the spin (and handedness) of the field. This equation defines a spacetime field $\phi_{\dots}(r)$ which automatically satisfies the field equations for a massless particle. Thus the holomorphicity constraint of twistor fields encodes all the messy field equations of a massless particle, at least for a linear field in flat space, or the weak energy limit of an Einstein field.

Geometrically the point r in spacetime is a $\mathbb{CP}_1$ line (which is a Riemann sphere) in twistor space. This line must cut through the region where $f(Z)$ is defined. $f(Z)$ is in general not defined everywhere and has singular places (indeed, we surround these singular regions to evaluate the contour integral). To be more mathematically precise, a twistor wave function is a *cohomology* element. To understand this, consider a collection of open neighborhoods of the region of twistor space in which we are interested. The twistor function must then be defined on the *intersection* of pairs of these open sets. This means that it is an element of the first sheaf cohomology. I shall not go into detail about this, but "sheaf cohomology" is a good buzzword to use!

Recall now that what we really want, in analogy with QFT, is a way of separating the positive and negative frequency parts of field amplitudes. If a twistor function defined on $\mathbb{PN}$ extends (as an element of the first cohomology) to the top half of twistor space $\mathbb{PT}^+$, it is of positive frequency. If it extends to the bottom half $\mathbb{PT}^-$, it is of negative frequency. Thus twistor space captures the notions of positive and negative frequency.

This splitting allows us to do quantum physics in twistor space. Andrew Hodges (1982, 1985, 1990) has developed an approach to QFT using twistor diagrams, which are analogous to Feynman diagrams in spacetime. Using these he has come up with some very novel ways of regularizing QFT. These are schemes that one wouldn't think of adopting in the normal spacetime approach but which are very natural in the twistor picture. A new angle arising, originally, from an idea due to Michael Singer (Hodges, Penrose, and Singer 1989) has also been stimulated by *conformal field theory* (CFT). Stephen made some very derogatory remarks about string theory in his first lecture, but I think that CFT, which is the field theory on the world-sheet of string theory, is a very beautiful (though not altogether physical) theory. It is defined on arbitrary Riemann surfaces (of which the Riemann sphere is the simplest example, but which include all the one-complex-dimensional manifolds such as tori and "pretzels"). For twistors we need to generalize CFT to manifolds with three complex dimensions whose boundaries are copies of $\mathbb{PN}$ (i.e., spaces of light rays in spacetime). The work in this area is progressing but has not moved very far yet.

TWISTORS FOR CURVED SPACES

All we have done so far relates only to flat spacetime, but we know that spacetime is curved; we need a theory of twistors that applies to curved spacetime and reproduces Einstein's equations in some natural way.

If the manifold of spacetime is conformally flat (or in other words, if its Weyl tensor is zero), there is no problem with describing this space with twistors, as twistor theory is basically conformally invariant. There are also twistor ideas that work for various conformally nonflat spacetimes, such as the definition of quasi-local mass (Penrose 1982; cf. Tod 1990), and the Woodhouse-Mason (1988; cf. also Fletcher and Woodhouse 1990) construction for stationary axisymmetric vacuums (based on Ward's 1977 construction for anti-self-dual Yang-Mills fields on flat spacetime; cf. also Ward 1983), which is part of a very general twistor approach to integrable systems (see the forthcoming book by Mason and Woodhouse 1996).

However, we should like to be able to cope with more general spacetimes. For a complexified (or "Euclideanized") spacetime M with anti-self-dual Weyl tensor (i.e., the self-dual half of the Weyl tensor is zero) there is a construction—the so-called nonlinear graviton construction—that fully addresses this problem (Penrose 1976). To see how this works, we take a part of twistor space consisting of a tubular neighborhood of a line, or something similar (say the top half or positive frequency part $\mathbb{PT}^+$), and cut it into two or more bits.

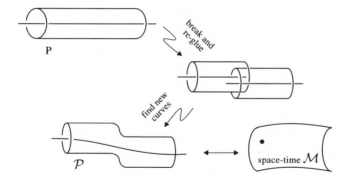

Figure 6.5 The nonlinear graviton construction.

It is then glued back together but with the bits shifted relative to each other. In general, the straight lines in the original space P would be broken in the new space $\mathcal{P}$. However, we can look for new holomorphic curves to replace the original (now broken) straight lines, providing curves that are smoothly joined together. Provided that the deformation $\mathcal{P}$ of P is not too great, the holomorphic curves that are obtained in this way—belonging to the same topological family as the original lines—form a four-dimensional family. The space whose points represent these holomorphic curves is our anti-self-dual (complex) "spacetime" $\mathcal{M}$ (fig. 6.5). Now we can encode the Einstein vacuum equations (Ricci-flatness) as the condition that $\mathcal{P}$ be a holomorphic fibration over a projective line $\mathbb{CP}_1$ (together with some other mild conditions). All this can be achieved by expressing the deformation $\mathcal{P}$ of P as being given in terms of *free* holomorphic functions, and in principle all the information of the curved spacetime $\mathcal{M}$ is encoded in these functions (although the finding of the required holomorphic curves in $\mathcal{P}$ can be a difficult matter).

We really want to solve the *full* Einstein equations (as the last construction only solves a reduced problem in which half the Weyl tensor is zero), but the problem is clearly difficult and has defeated many attempts over the last twenty years. In the last few years, however, I have been trying a new approach (cf. Penrose 1992). Although I have no solution to the problem as yet, it looks to be the most promising way forward so far. There indeed appears to be a deep relation between twistors and Einstein's equations. This is indicated by two observations:

1. The vacuum Einstein equations $R_{ab} = 0$ are also the consistency conditions for massless fields of helicity $s = \frac{3}{2}$ (when that field is given in terms of a potential).
2. In flat spacetime M the space of charges of an $s = \frac{3}{2}$ field is exactly twistor space.

The program to be carried through is then roughly the following: given a Ricci-flat spacetime (i.e., $R_{ab} = 0$), one has to find the space of charges for $s = \frac{3}{2}$ fields in it (which is not an easy task). This would then be the twistor space of the Ricci-flat spacetime. The second step is to find how to construct such twistor spaces using free holomorphic functions and, finally, to reconstruct the original spacetime manifold from this twistor space in each case.

We don't expect this twistor space to be linear, as it must give a curved structure when we reconstruct spacetime. Also, the construction must be highly nonlocal in a subtle way, as both the charge and potential of an $s = \frac{3}{2}$ field are nonlocal. This would be expected to help in explaining nonlocal physics such as the EPR experiments discussed in my last lecture (chapter 4)—these experiments imply that objects in distant regions in spacetime can somehow be "entangled" with one another.

TWISTOR COSMOLOGY

I want to finish by making a remark about cosmology and twistors—although it will be rather tentative. I have said that the Weyl curvature tensor has to be zero at past singularities, and that the spacetime is close to being conformally flat there. This means that the initial state has a very simple twistor description. This description gets more and more complicated as time proceeds, and Weyl curvature gets more pervasive. This type of behavior is consistent with the observed time asymmetry in the geometry of the universe.

From the point of view of the complex-holomorphic ideology of twistor theory, a big bang with $k < 0$, leading to an open universe, is to be preferred (Stephen prefers a closed one). The reason is that only in a $k < 0$ universe is the symmetry group of the initial singularity a holomorphic group, namely just the Möbius group of holomorphic self-transformations of the Riemann sphere $\mathbb{CP}_1$ (i.e., the restricted Lorentz group). This is the same group that started twistor theory off in the first place—so, for twistor-ideological reasons, I certainly prefer $k < 0$. Since this is based only on ideology I can, of course, withdraw it in the future if I am wrong and the universe is, in fact, found to be closed!

QUESTIONS AND ANSWERS

Question: What is the physical significance of the helicity $\frac{3}{2}$ state?

Answer: The spin $\frac{3}{2}$ of this approach is no actual physical field, but rather an auxiliary field for the definition of twistors. I don't think of it as the field of a particle that one might discover. On the other hand, from the point of view of supersymmetry, it would be the superpartner of the graviton.

Question: Where does the time-asymmetrical **R**-process you talked about last time appear in the twistor point of view?

Answer: You have to realize that twistor theory is a very conservative theory and doesn't say anything about that, yet. I would very much like to see the time asymmetry appear in twistor theory, but at present I don't know how this is going to come about. However, if one carries through the whole program it certainly should appear, maybe in a vaguely similar way to the right/left asymmetry. Also, Andrew Hodges's approach to the regularization scheme technically introduces a time asymmetry, but the dust has not settled on this, yet.

Question: Which nonlinear QFT might be most amenable to twistor theory?

Answer: So far mainly the standard model has been analyzed (in the context of twistor diagrams).

Question: String theory explicitly predicts the spectrum of particles. Where does this appear in twistor theory?

Answer: I don't know how the particle spectrum could finally emerge, although there have been some ideas on this. However, I am pleased to learn that string theory "explicitly predicts the spectrum of particles." My view is that until we understand GR in the twistor framework, we shall not be able to solve this problem, as masses are tied up with GR. But in a sense this is the string theory point of view as well.

Question: What is the twistor point of view on continuity/discontinuity?

Answer: Another early motivation for twistor theory was the theory of spin networks, where one strives to build up space from discrete combinatorial quantum rules. One can try to construct twistor theory out of discrete things as well. However, over the years the trend has moved away to holomorphic rather than combinatorial methods, but this doesn't mean that the discrete point of view is inferior. Maybe there is a deep connection between discrete concepts and holomorphic concepts, but this hasn't emerged in any clear way yet.

CHAPTER SEVEN

The Debate

S. W. Hawking and R. Penrose

STEPHEN HAWKING

THESE LECTURES HAVE SHOWN very clearly the difference between Roger and me. He's a Platonist and I'm a positivist. He's worried that Schrödinger's cat is in a quantum state, where it is half alive and half dead. He feels that can't correspond to reality. But that doesn't bother me. I don't demand that a theory correspond to reality because I don't know what it is. Reality is not a quality you can test with litmus paper. All I'm concerned with is that the theory should predict the results of measurements. Quantum theory does this very successfully. It predicts that the result of an observation is either that the cat is alive or that it is dead. It is like you can't be slightly pregnant: you either are or you aren't.

The reason that people like Roger, not to mention the animal liberation front, object to Schrödinger's cat is that it seems absurd to represent the state as $\frac{1}{\sqrt{2}}(\text{cat}_{alive} + \text{cat}_{dead})$. Why not $\frac{1}{\sqrt{2}}(\text{cat}_{alive} - \text{cat}_{dead})$. Another way of saying it is that there doesn't seem to be any interference between cat_{dead} and cat_{alive}. You can get interference between particles going through different slits, because one can isolate them reasonably well from the environment that one doesn't measure. But one can't isolate something as large as a cat from ordinary intermolecular forces carried by the electromagnetic field. One doesn't need quantum gravity to explain Schrödinger's cat or the operation of the brain. It is a red herring.

I was not seriously suggesting that cosmological event horizons are the reason that Schrödinger's cat appears to be a classical animal that is either dead or alive but not a combination of the two. As I have said, it would be difficult enough to isolate the cat from the rest of the room, so that one doesn't need to worry about the far reaches of the universe. All I was saying was that even if we could observe the fluctuations in the microwave background with great accuracy, they would appear to have a classical statistical distribution. We could not detect any quantum state properties like interference or correlations between the fluctuations in different modes. When we talk about the whole universe we don't have an external environment like we did in the case of Schrödinger's cat, but we still get decoherence and classical behavior because we can't see the whole universe.

Roger questions my use of Euclidean methods. In particular he objects to the pictures I drew of a Euclidean geometry joined to a Lorentzian one. As he rightly says, this is possible only for very special cases: a general Lorentzian spacetime will not have a section in the complexified manifold on which the metric is real and positive definite or Euclidean. However, this is to misunderstand the Euclidean path integral approach even for ordinary nongravitational fields. Let us take the Yang-Mills case, which is well understood. Here one starts with a path integral $e^{i \text{ action}}$ over all Yang-Mills connections in Minkowski space. This integral oscillates and does not converge. To get a better-behaved path integral one does a Wick rotation to Euclidean space by introducing the imaginary time coordinate $\tau = -it$. The integrand then becomes $e^{-\text{Euclidean action}}$ and one does the path integral over all real connections in Euclidean space. A connection that is real in Euclidean space will in general not also be real in Minkowski space. But that doesn't matter. The idea is that the path integral over all real connections in Euclidean space is equivalent in the sense of contour integrals to a path integral over all real connections in Minkowski space. As in the case of quantum gravity, one can evaluate the Yang-Mills path integral by saddle point methods. Here the saddle point solutions are the Yang-Mills instantons which Roger and the twistor program have done so much to classify. The Yang-Mills instantons are real in Euclidean space. But they are complex in Minkowski space. This doesn't matter. They still give the rates for physical processes like electroweak baryon generation.

The situation for quantum gravity is similar. Here one can take the path integral to be over positive definite or Euclidean metrics rather than over Lorentzian ones. Indeed, it is necessary to do this if one is to allow the gravitational field to have different topologies. One can have a Lorentzian metric only on a manifold with zero Euler number. But as we have seen, the interesting quantum gravitational effects like intrinsic entropy appear precisely from spacetime manifolds with nonzero Euler number that don't admit Lorentzian metrics. There is a problem in that the Euclidean action for gravity is not bounded below, so it looks like the path integral wouldn't converge. However, one can cure this by integrating the conformal factor over a complex contour. This is a fudge, but I think this behavior is related to the gauge freedom, and will cancel when we know how to do the path integral properly. This problem arises for a physical reason: the potential energy of gravity is negative because gravity is attractive. Thus, it will appear in some form in any theory of quantum gravity. It will be there in string theory if it ever gets that far. So far, its performance has been pretty pathetic: string theory cannot even describe the structure of the Sun, let alone black holes.

After taking that side swipe at string theory, let me return to the Euclidean approach and the no-boundary condition. Although the path integral is to be taken over positive definite real metrics, the saddle point may well be a complex metric. This will happen in cosmology when the three-surface Σ is larger than some very small size. Although I described the metric as half a Euclidean four-sphere joined to a Lorentzian metric, this was only approximate. The actual saddle point metric will be complex. This may upset a Platonist like Roger but it is fine for a positivist like me. One doesn't observe the saddle point metric. All one can observe is the wave function calculated from it, and this corresponds to a real Lorentzian metric. I'm a bit surprised at Roger objecting to my use of Euclidean and complex spacetime. He uses complex spacetime in his twistor program. Indeed, it was Roger's comments about positive frequency being holomorphic that led me to develop the Euclidean quantum gravity program. I would claim that this program has made two observationally testable predictions. How many predictions has string theory or the twistor program made?

Roger feels that observation or measurement through the **R** process, the collapse of the wave function, introduces CPT violation into physics. He sees such violations at work in at least two situations: cosmology and black holes. I agree that we may introduce time asymmetry in the way we ask questions about observations. But I totally reject the idea that there is some physical process that corresponds to the reduction of the wave function or that this has anything to do with quantum gravity or consciousness. That sounds like magic to me, not science.

I have already explained in my lectures why I think that the no-boundary proposal can explain the observed arrow of time in cosmology without any CPT violation. I will now explain why, unlike Roger, I don't think black holes involve any time asymmetry either. In classical general relativity, a black hole is defined as a region that objects can fall into but nothing can get out of. Why, one might ask, aren't there also white holes, regions that objects can come out of but nothing can fall into? My answer is that although black and white holes are very different in classical theory, they are the same thing in quantum theory. Quantum theory removes the distinction between black and white holes: black holes can emit, and presumably white holes can absorb. I would suggest that we refer to a region as a black hole when it is large and classical and not emitting much. On the other hand, a small hole that is sending out large amounts of quantum radiation is just as we would expect a white hole to behave.

I shall illustrate how black and white holes are the same using the thought experiment that Roger has referred to. One places a certain amount of energy in a very large box with perfectly reflecting walls. This energy can distribute itself in various ways among the possible states in the box. Two possible situations correspond to the overwhelming majority of the states. They are a boxed filled with thermal radiation or a black hole in equilibrium with thermal radiation. Which situation has the higher number of microscopic states depends on the size of the box and the amount of energy in it. But one could choose these parameters so the two situations corresponded to

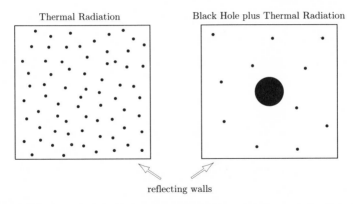

Figure 7.1 A box containing a fixed energy will contain either just thermal radiation, or else a black hole in equilibrium with the thermal radiation.

roughly equal numbers of microscopic states. One would then expect the box to fluctuate between the situations. At some times the box will contain just thermal radiation. Then at other times thermal fluctuations in the radiation will mean that a very large number of particles are in a small region, and a black hole would form (fig. 7.1). At yet further times the radiation from the black hole would fluctuate upward or the absorbtion would fluctuate downward, and the black hole would evaporate and disappear. Thus the system in the box would wander ergodically through its phase space: sometimes there would be a black hole present and at other times there would not (fig. 7.2).

Roger and I agree that the box would behave in the way I have described. But we disagree on two points. First, Roger believes that phase-space volume and information will be lost during this cycle of the appearance and disappearance of black holes; and second, that the process will not be time symmetric. On the first point, Roger seems to feel that the black hole no-hair theorems imply loss of phase-space volume because many different configurations of the collapsing particles produce the same black hole. He suggests that the **R** process, the collapse of the wave function, introduces compensating gain in phase-space volume. It is not clear to me how this **R** process is supposed to come about. There are no observers in the box, and I'm not

History of Box

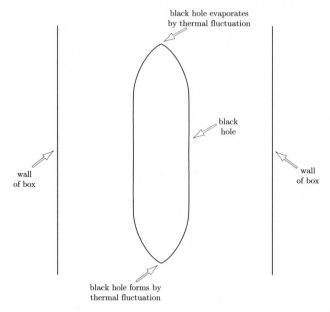

Figure 7.2 A black hole appears and disappears by thermal fluctuations.

sympathetic to claims that it is spontaneous, unless one can suggest a way of calculating it. Otherwise, it is just magic. Anyway, I don't agree that there is loss of phase-space volume. If you say that black holes have a number of states equal to $e^{\frac{1}{4}A}$, there's no loss of phase-space volume. And there's no information in a system like the box that can be in any state. So there's no loss of information.

To turn to our second disagreement, I believe that the appearance and disappearance of black holes will be time symmetric. That is, if you take a film of the box and run it backward it will look the same. In one direction of time, one will have black holes appearing and disappearing. In the other direction, you will have white holes—the time reverse of black holes—appearing and disappearing. These two pictures can be the same if white holes are the same as black holes. Thus, there is no need to invoke CPT violation because of the behavior of this box (fig. 7.3).

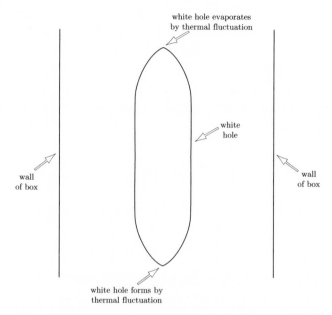

Figure 7.3 A white hole appears and disappears by thermal fluctuations.

Initially both Roger and Don Page rejected my suggestion that the formation and evaporation of black holes in the box was time symmetric. However, Don has now come around to agreeing with me. I'm waiting for Roger to do the same.

ROGER PENROSE REPLIES

Let me first say that I believe that there is more agreement than disagreement between us. However, there are certain (fundamental) points where we disagree and I want to concentrate on those in the following.

Cats and the Like

Whatever "reality" may be, one has to explain how one perceives the world to be. QM does not do this and one must incorporate something additional into QM—something not contained in the standard rules of QM. In particular, it seems to me that Stephen hasn't quite taken in my remarks about the problem with the cat. The problem is not that the loss of information implies that the system must be described by a density matrix, but that the two density matrices

$$D = \frac{1}{4}(|\text{live}\rangle + |\text{dead}\rangle)(\langle\text{live}| + \langle\text{dead}|)$$

$$+ \frac{1}{4}(|\text{live}\rangle - |\text{dead}\rangle)(\langle\text{live}| - \langle\text{dead}|) \qquad (7.1)$$

and

$$D = \frac{1}{2}|\text{live}\rangle\langle\text{live}| + \frac{1}{2}|\text{dead}\rangle\langle\text{dead}|, \qquad (7.2)$$

for example, are equal. Therefore we have to solve the problem of why we do perceive either a live cat or a dead cat, but never a superposition. I think philosophy is important in these matters, but it doesn't answer the question.

It seems to me that in order to explain how we perceive the world to be, within the framework of QM, we shall need to have one (or both) of the following:

(A) A theory of experience.
(B) A theory of real physical behavior.

In fact, bringing the observer into play, the corresponding state vectors (in case 7.1 above) would each have the form

$$\frac{1}{2}(|\text{live}\rangle \pm |\text{dead}\rangle)(|\text{observer sees live cat}\rangle \pm |\text{observer sees dead cat}\rangle).$$
$$(7.3)$$

Then the first alternative (A) would have to be to rule out the possibility of the superposition in the second factor, as this state of perception would not be allowed. The requirement for (B), on the other hand, would rule out the superposition in the first factor. In my own picture, these large-scale superpositions are unstable, and they must rapidly decay (spontaneously) into one or the other stable states |live⟩ or |dead⟩. I believe that Stephen must be an A-supporter [SWH: No], because he isn't a B-supporter. I am a strong B-supporter, as I believe that (A) is a dangerous view to adopt, which leads into all sorts of troubles. In particular, an A-supporter needs a theory of the mind or the brain or something like that. I am surprised that Stephen seems to be neither an A- nor a B-supporter; I am looking forward to his commenting about this.

The Wick Rotation

This is a useful tool in QFT. One replaces t by it by means of a rotation of the time axis. This translates Minkowski space into Euclidean space. Its usefulness stems from the fact that certain expressions (such as path integrals) are better defined in the Euclidean theory. Wick rotation is a well-controlled tool in QFT, at least as long as one applies it to flat (or stationary) spacetime.

Stephen's idea of applying the "Wick rotation" to the space of Lorentzian metrics (to obtain the space of Euclidean metrics) is certainly very interesting and ingenious, but it is a very different procedure from that of applying a Wick rotation in QFT. It is really a "Wick rotation" on a different level.

The NBP is a very nice proposal and certainly seems to be related to the Weyl curvature hypothesis. However, from my point of view NBP is very far from being an explanation of the fact that past singularities have small Weyl curvature whereas future singularities have large Weyl curvature. This is what we observe in our universe, and I believe that on the observational side Stephen agrees with me.

Phase-Space Loss

I think Stephen and I agree that there is information loss in a black hole, but disagree about the loss of phase space in a black hole. Stephen has claimed that the **R**-process is mere magic but not physics.

I obviously don't agree with this; I think I have explained in my second lecture why this is reasonable and have made a definite proposal for the rate at which the reduction of the state should take place, namely in a time

$$T \sim \frac{\hbar}{E}. \tag{7.4}$$

I also think that his diagram of the black hole is very misleading. He should have drawn the Carter diagram, and then it is obviously not time symmetric. He and I seem to agree anyway that information is lost, but I also believe that the phase-space volume is reduced. Furthermore, if the whole scheme were time symmetrical, we should be allowed to have white holes, which are regions out of which lots of things can come, and that would be at least in disagreement with the Weyl curvature hypothesis, with the second law of thermodynamics, and probably also with observation. This question is very much tied up with what type of singularities "quantum gravity" will allow. In my view, it is necessary that that theory be time asymmetric in its implications.

STEPHEN HAWKING

Roger is worried about Schrödinger's poor cat. Such a thought experiment would not be politically correct nowadays. Roger is concerned because a density matrix that has cat_{alive} and cat_{dead} with equal probabilities also has $cat_{alive} + cat_{dead}$ and $cat_{alive} - cat_{dead}$ with equal probabilities. So why do we observe either cat_{alive} or cat_{dead}? Why don't we observe either $cat_{alive} + cat_{dead}$ or $cat_{alive} - cat_{dead}$? What is it that picks the *alive* and *dead* axes for our observations rather than *alive + dead* and *alive − dead*. The first point I would make is that one gets this ambiguity in the eigenstates of the density matrix only when the eigenvalues are exactly equal. If the probabilities of being *alive* or *dead* were slightly different, there would be no ambiguity in the eigenstates. One basis would be distinguished by being eigenvectors of the density matrix. So why does nature choose to make the density matrix diagonal in the *alive/dead* basis rather than in the *alive + dead / alive − dead* basis? The answer is that the cat_{alive} and cat$_{dead}$ states differ on a macroscopic level by things like the position of the bullet or the wound on the cat. When you trace out over the things you don't observe, like the disturbance in the air molecules, the matrix element of any observable between the cat$_{alive}$ and cat$_{dead}$ states will average out to zero. That is why one observes the cat either dead *or* alive and not a linear combination of the two. This is just ordinary quantum mechanics. One doesn't need a new theory of measurement, and one certainly doesn't need quantum gravity.

Let's get back to quantum gravity. Roger seems to accept that the no-boundary proposal can explain the low Weyl tensor in the early universe. However, he questions whether it can account for the high Weyl tensor that is expected in gravitational collapse in black holes and the collapse of the whole universe. I think this again is based on a misconception about the no-boundary proposal. Roger would presumably agree that there are Lorentzian solutions that start in the early universe being almost smooth and develop into highly irregular metrics in gravitational collapse. One can join these Lorentzian metrics to half a Euclidean four-sphere in the early universe. This will give an approximate saddle-point metric for the wave function of a highly distorted three-geometry in the collapse (fig. 7.4). Of course, as I said earlier, the exact saddle point metric will be complex and won't be either Euclidean or Lorentzian. Nevertheless, to a good approximation one can divide it into nearly Euclidean and Lorentzian regions as I have described. The Euclidean region will be only slightly different from half the round four-sphere. Thus its action will be only slightly higher than half the round four-sphere, which corresponds to a homogeneous and isotropic universe. The Lorentzian part of the solution will be very different from a homogeneous and isotropic solution. However, the action of this Lorentzian part merely changes the phase of the wave function and does not affect the amplitude. This is given by the action of the Euclidean part and will be almost independent of how distorted the three-geometry is. Thus all three-geometries are equally probable in gravitational collapse, and one will typically have a very irregular metric with a lot of Weyl curvature. I hope this will convince Roger, and everyone else for that matter, that the no-boundary proposal can explain both

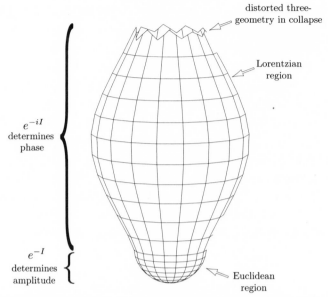

distorted three-
geometry in collapse

Lorentzian
region

e^{-iI}
determines
phase

e^{-I}
determines
amplitude

Euclidean
region

Figure 7.4 In the tunneling to collapsed three-geometry, the Euclidean section determines the amplitude of the wavefunction for the three-geometry, while the Lorentzian section determines its phase.

why the early universe was smooth and why gravitational collapse will be irregular.

My last points are about the black hole in a box thought experiment. Roger still seems to be claiming that there is loss of phase-space volume because many different configurations can collapse to form the same black hole. But the whole point of black hole thermodynamics was to avoid such a loss of phase space. One attributes an entropy to black holes precisely because they can be formed in e^S ways. When they evaporate in a time-symmetric way they send out radiation in e^S ways. Thus there is no loss of phase-space volume and no need to invoke the **R** process to compensate. Just as well: I believe in gravitational collapse, but not in the collapse of the wave function.

My final point is about my claim that black and white holes are the same. Roger objects that the Carter-Penrose diagrams are very different (fig. 7.5). I agree that they are different but would say that

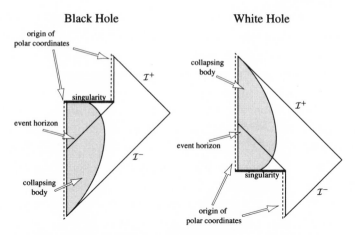

Figure 7.5 The Carter-Penrose diagrams for black and white holes.

they are only a classical picture. In quantum theory, I would claim that black and white holes are the same to an outside observer. But, Roger might object, what about someone who falls into a hole? Won't he or she see the black hole Carter-Penrose diagram? I think this argument falls into the trap of assuming that there is a single metric for spacetime, as there is in classical theory. In quantum theory, on the other hand, one has to do a path integral over all possible metrics. There will be different saddle point metrics for different questions. In particular, the saddle point metrics for the questions that outside observers ask will be different from the saddle point metric for an infalling observer. One could also imagine that the black hole could emit an observer. The probability is small but it is possible. Presumably the saddle point metric for such an observer would correspond to the white hole Carter-Penrose diagram. Thus my claim that black and white holes are the same is consistent. It is the only natural way to make quantum gravity CPT invariant.

ROGER PENROSE REPLIES

Let me come back to Stephen's remark about the cat problem. In fact, the equality of the eigenvalues is irrelevant. It has been shown recently (Hughston et al. 1993) that for any density matrix (even with completely distinct eigenvalues), for all the many different ways in which it can be written as a probability mixture of (not necessarily orthogonal) states, there is a measurement that one can, in principle, perform on the "unknown part of the state-vector" which gives that particular probability mixture the interpretation of the density matrix for the "known part." Moreover, as far as the effect of the environment is concerned, it may be remarked that even though the off-diagonal terms might be small, the effect of them on the eigenvectors might be large. Furthermore, Stephen has also mentioned bullets, etc. This doesn't really address the problem, because we have the same problem for the system of "cat+bullet" as we had before for the "cat" alone. I think this question of "reality" is the fundamental difference between Stephen and me, and it relates to the other problems—for example, to the problem of whether white holes and black holes are the same. It all really boils down to the fact that on the macroscopic level we perceive only one spacetime. Thus, it seems to me, one has to support either (A) or (B)—I don't feel Stephen has addressed this point.

Black and white holes might be very similar for very small holes. A small black hole would be emitting lots of radiation and so might look like a white hole. Presumably a small white hole could also absorb a large amount of radiation. But on the macroscopic level this identification seems to me to be inappropriate; I believe something else has to come in.

QM has only been around for seventy-five years. This is not very long if one compares it, for example, with Newton's theory of gravity. Therefore it wouldn't surprise me if QM will have to be modified for very macroscopic objects.

At the beginning of this debate Stephen said that he thinks that he is a positivist, whereas I am a Platonist. I am happy with him being a positivist, but I think that the crucial point here is, rather, that I am a realist. Also, if one compares this debate with the famous debate of Bohr and Einstein, some seventy years ago, I should think that Stephen plays the role of Bohr, whereas I play Einstein's role! For Einstein argued that there should exist something like a real world, not necessarily represented by a wave function, whereas Bohr stressed that the wave function doesn't describe a "real" microworld but only "knowledge" that is useful for making predictions.

Bohr was perceived to have won the argument. In fact, according to the recent biography of Einstein by Pais (1994), Einstein might as well have gone fishing from 1925 onward. Indeed, it is true that he didn't make many big advances, even though his penetrating criticisms were very useful. I believe that the reason why Einstein didn't continue to make big advances in quantum theory was that a crucial ingredient was missing from QT. This missing ingredient was Stephen's discovery, fifty years later, of black hole radiation. It is this information loss, connected with black hole radiation, which provides the new twist.

QUESTIONS AND ANSWERS

Gary Horowitz (Remark): There have been a few disparaging remarks about string theory. Even though they have been disparaging, the large number of them at least seems to indicate that string theory is quite important! Some of these remarks have been misleading, some quite simply wrong. First of all, string theory reduces in the weak field limit to GR and thus implies everything GR implies. It also might give a better understanding of what happens at the singularity, and in fact some of the uncontrollable divergencies seem to be solved by string theory. I am certainly not claiming that string theory has overcome all its problems, but it seems still to be a very promising route.

Question: A confused question, again about the cat.

Answer: Roger Penrose explains the cat problem again.

Question: Could Roger Penrose comment on the approach of decoherent histories? It has been shown that there is very good decoherence due to an external environment; however, it is not (yet) quite understood how decoherence would work internally. Maybe this is related to the fact that decoherence might be related to properties of the spacetime?

Answer (Penrose): In the decoherent histories program, something equivalent to the **R** operation is part of the scheme. So it is different from usual QM, but nevertheless it is also something different from my approach. However, it is interesting to hear that there might be a link to spacetime structure. I think my approach is less different from the consistent histories approach than from Stephen's with regard to the time-asymmetry question.

Question: What about the entropy in the thought experiment of the black hole in the box? Would the time-reversed situation not violate the second law of thermodynamics?

Answer (Hawking): The box is in a state of maximal entropy. The system is moving ergodically among all possible states, so there is no violation.

Question: Could the mechanism of the quantum measurement be tested experimentally?

Answer (Penrose): It should be possible (in principle) to test it experimentally. Maybe one should try some Leggett-type experiment, having some large-scale superposition. The trouble with these kinds of experiments is that the decoherence effects due to the environment are usually much larger than the effects one would like to measure. Thus one has to isolate the system very well indeed. As far as I know there are as yet no suggestions to test this idea in detail, but it would be certainly very interesting indeed.

Question: In an inflationary model of the universe the mass of the universe must be very well balanced between an expanding and a contracting universe. Only 10% of the mass necessary for this balance has been seen so far and the search for the remaining mass reminds me somehow of the search for the "ether" around the turn of the century. Would you like to comment on that?

Answer (Penrose): I am reasonably happy with a Hubble constant within the present range of values, and 10% of the critical mass would be fine for me. I have never been particularly happy with inflationary models anyway. But I think Stephen wants the universe to be closed, as part of the NBP. [SWH: Yes!]

Answer (Hawking): The Hubble constant might be less than claimed. It decreased by a factor of ten in the last fifty years, and I don't see why it shouldn't decrease by another factor of two. This would reduce the necessary mass to be found.

References

Aharonov, Y., Bergmann, P., and Lebowitz, J. L. 1964. Time symmetry in the quantum process of measurement. In *Quantum Theory and Measurement*, ed. J. A. Wheeler and W. H. Zurek. Princeton University Press, Princeton, 1983. Originally in *Phys. Rev.* **134B**, 1410–16.

Bekenstein, J. 1973. Black holes and entropy. *Phys. Rev.* **D7**, 2333–46.

Carter, B. 1971. Axisymmetric black hole has only two degrees of freedom. *Phys. Rev. Lett.* **26**, 331–333.

Diósi, L. 1989. Models for universal reduction of macroscopic quantum fluctuations. *Phys. Rev.* **A40**, 1165–74.

Fletcher, J., and Woodhouse, N. M. J. 1990. Twistor characterization of stationary axisymmetric solutions of Einstein's equations. In *Twistors in Mathematics and Physics*, ed. T. N. Bailey and R. J. Baston. LMS Lecture Notes Series 156. Cambridge University Press, Cambridge, U.K.

Gell-Mann, M., and Hartle, J. B. 1990. In *Complexity, Entropy, and the Physics of Information*. SFI Studies in the Science of Complexity, vol. 8, ed. W. Zurek. Addison-Wesley, Reading, Mass.

Geroch, R. 1970. Domain of dependence. *J. Math. Phys.* **11**, 437–449.

Geroch, R., Kronheimer, E. H., and Penrose, R. 1972. Ideal points in space-time. *Proc. Roy. Soc. London* **A347**, 545–567.

Ghirardi, G. C., Grassi, R., and Rimini, A. 1990. Continuous-spontaneous-reduction model involving gravity. *Phys. Rev.* **A42**, 1057–64.

Gibbons, G. W. 1972. The time-symmetric initial value problem for black holes. *Comm. Math. Phys.* **27**, 87–102.

Griffiths, R. 1984. Consistent histories and the interpretation of quantum mechanics. *J. Stat. Phys.* **36**, 219–272.

Hartle, J. B., and Hawking, S. W. 1983. Wave function of the universe. *Phys. Rev.* **D28**, 2960–2975.

Hawking, S. W. 1965. Occurrence of singularities in open universes. *Phys. Rev. Lett.* **15**, 689–690.

Hawking, S. W. 1972. Black holes in general relativity. *Comm. Math. Phys.* **25**, 152–166.

Hawking, S. W. 1975. Particle creation by black holes. *Comm. Math. Phys.* **43**, 199–220.

Hawking, S. W., and Penrose, R. 1970. The singularities of gravitational collapse and cosmology. *Proc. Roy. Soc. London* **A314**, 529–48.

Hodges, A. P. 1982. Twistor diagrams. *Physica* **114A**, 157–75.

Hodges, A. P. 1985. A twistor approach to the regularization of divergences. *Proc. Roy. Soc. London* **A397**, 341–74. Also, Mass eigenstates in twistor theory, ibid., 375–96.

Hodges, A. P. 1990. Twistor diagrams and Feynman diagrams. In *Twistors in Mathematics and Physics*, ed. T. N. Bailey and R. J. Baston. LMS Lecture Notes Series 156. Cambridge University Press, Cambridge, U.K.

Hodges, A. P., Penrose, R., and Singer, M. A. 1989. A twistor conformal field theory for four space-time dimensions. *Phys. Lett.* **B216**, 48–52.

Huggett, S. A., and Tod, K. P. 1985. *An Introduction to Twistor Theory*. London Math. Soc. student texts. LMS publication, Cambridge University Press, New York.

Hughston, L. P., Jozsa, R., and Wooters, W. K. 1993. A complete classification of quantum ensembles having a given density matrix. *Phys. Lett.* **A183**, 14–18.

Israel, W. 1967. Event horizons in static vacuum space-times. *Phys. Rev.* **164**, 1776–1779.

Majorana, E. 1932. Atomi orientati in campo magnetico variabile. *Nuovo Cimento* **9**, 43–50.

Mason, L. J., and Woodhouse, N. M. J. 1996. *Integrable Systems and Twistor Theory* (tentative). Oxford University Press, Oxford (forthcoming).

Newman, R. P. A. C. 1993. On the structure of conformal singularities in classical general relativity. *Proc. Roy. Soc. London* **A443**, 473–92; II, Evolution equations and a conjecture of K. P. Tod, ibid., 493–515.

Omnès, R. 1992. Consistent interpretations of quantum mechanics. *Rev. Mod. Phys.* **64**, 339–82.

Oppenheimer, J. R., and Snyder, H. 1939. On continued gravitational contraction. *Phys. Rev.* **56**, 455–59.

Pais, A. 1994. *Einstein Lived Here*. Oxford University Press, Oxford.

Penrose, R. 1965. Gravitational collapse and space-time singularities. *Phys. Rev. Lett.* **14**, 57–59.

Penrose, R. 1973. Naked singularities. *Ann. N.Y. Acad. Sci.* **224**, 125–134.

Penrose, R. 1976. Non-Linear gravitons and curved twistor theory. *Gen. Rev. Grav.* **7**, 31–52.

Penrose, R. 1978. Singularities of space-time. In *Theoretical Principles in Astrophysics and Relativity*, ed. N. R. Liebowitz, W. H. Reid, and P. O. Vandervoort. University of Chicago Press, Chicago.

Penrose, R. 1979. Singularities and time-asymmetry. In *General Relativity: An Einstein Centenary*, ed. S. W. Hawking and W. Israel. Cambridge University Press, Cambridge, U.K.

Penrose, R. 1982. Quasi-local mass and angular momentum in general relativity. *Proc. Roy. Soc. London* **A381**, 53–63.

Penrose, R. 1986. On the origins of twistor theory. In *Gravitation and Geometry* (I. Robinson Festschrift volume), ed. W. Rindler and A. Trautman. Bibliopolis, Naples.

Penrose, R. 1992. Twistors as spin 3/2 charges. In *Gravitation and Modern Cosmology* (P. G. Bergmann's 75th Birthday volume), ed. A. Zichichi, N. de Sabbata, and N. Sánchez. Plenum Press, New York.

Penrose, R. 1993. Gravity and quantum mechanics. In *General Relativity and*

Gravitation 1992. Proceedings of the Thirteenth International Conference on General Relativity and Gravitation held at Cordoba, Argentina, 28 June–4 July 1992. Part 1, Plenary Lectures, ed. R. J. Gleiser, C. N. Kozameh, and O. M. Moreschi. Institute of Physics Publication, Bristol and Philadelphia.

Penrose, R. 1994. *Shadows of the Mind: An Approach to the Missing Science of Consciousness*. Oxford University Press, Oxford.

Penrose, R., and Rindler, W. 1984. *Spinors and Space-Time*, vol. 1: *Two-Spinor Calculus and Relativistic Fields*. Cambridge University Press, Cambridge.

Penrose, R., and Rindler, W. 1986. *Spinors and Space-Time*, vol. 2: *Spinor and Twistor Methods in Space-Time Geometry*. Cambridge University Press, Cambridge.

Rindler, W. 1977. *Essential Relativity*. Springer-Verlag, New York.

Robinson, D. C. 1975. Uniqueness of the Kerr black hole. *Phys. Rev. Lett.* **34**, 905–906.

Seifert, H.-J. 1971. The causal boundary of space-times. *J. Gen. Rel. and Grav.* **1**, 247–259.

Tod, K. P. 1990. Penrose's quasi-local mass. In *Twistors in Mathematics and Physics*, ed. T. N. Bailey and R. J. Baston. LMS Lecture Notes Series 156. Cambridge University Press, Cambridge, U.K.

Ward, R. S. 1977. On self-dual gauge fields. *Phys. Lett.* **61A**, 81–82.

Ward, R. S. 1983. Stationary and axi-symmetric spacetimes. *Gen. Rel. Grav.* **15**, 105–9.

Woodhouse, N. M. J., and Mason, L. J. 1988. The Geroch group and non-Hausdorff twistor spaces. *Nonlinearity* **1**, 73–114.